V/STOL AIRCRAFT PROGRAMS

Committee Print

REVIEW OF DEFENSE-RELATED VERTICAL AND SHORT TAKEOFF AND LANDING (V/STOL) AIRCRAFT PROGRAMS

A STAFF STUDY FOR THE COMMITTEE ON ARMED SERVICES HOUSE OF REPRESENTATIVES

NINETY-SIXTH CONGRESS

FIRST SESSION

Prepared By
DR. WALTER Z. COLLINGS
American Institute of Aeronautics and Astronautics
Congressional Science and Engineering Fellow
October 1977-Septmeber 1978

© Ross & Perry, Inc. 2001 All rights reserved.

No claim to U.S. government work contained throughout this book.

Protected under the Berne Convention. Published 2001

Printed in The United States of America
Ross & Perry, Inc. Publishers
717 Second St., N.E., Suite 200
Washington, D.C. 20002
Telephone (202) 675-8300
Facsimile (202) 675-8400
info@RossPerry.com

SAN 253-8555

Government Reprints Press Edition 2001

Government Reprints Press is an Imprint of Ross & Perry, Inc.

Library of Congress Control Number: 2001093144
http://www.GPOreprints.com

ISBN 1-931641-36-6

∞ The paper used in this publication meets the requirements for permanence established by the American National Standard for Information Sciences "Permanence of Paper for Printed Library Materials" (ANSI Z39.48-1984).

All rights reserved. No copyrighted part of this publication may be reproduced, stored in a retrieval system, or transmitted, in any form or by any means, electronic, photocopying, *recording*, or otherwise, without the prior written permission of the publisher.

HOUSE COMMITTEE ON ARMED SERVICES

NINETY-SIXTH CONGRESS

MELVIN PRICE, Illinois, *Chairman*

CHARLES E. BENNETT, Florida
SAMUEL S. STRATTON, New York
RICHARD H. ICHORD, Missouri
LUCIEN N. NEDZI, Michigan
CHARLES H. WILSON, California
RICHARD C. WHITE, Texas
BILL NICHOLS, Alabama
JACK BRINKLEY, Georgia
ROBERT H. (BOB) MOLLOHAN, West Virginia
DAN DANIEL, Virginia
G. V. (SONNY) MONTGOMERY, Mississippi
HAROLD RUNNELS, New Mexico
LES ASPIN, Wisconsin
RONALD V. DELLUMS, California
MENDEL J. DAVIS, South Carolina
PATRICIA SCHROEDER, Colorado
ABRAHAM KAZEN, JR., Texas
ANTONIO B. WON PAT, Guam
BOB CARR, Michigan
JIM LLOYD, California
LARRY McDONALD, Georgia
BOB STUMP, Arizona
VIC FAZIO, California
CLAUDE (BUDDY) LEACH, Louisiana
BEVERLY B. BYRON, Maryland
NICHOLAS MAVROULES, Massachusetts
JOE WYATT, JR., Texas
DON BAILEY, Pennsylvania

BOB WILSON, California
WILLIAM L. DICKINSON, Alabama
G. WILLIAM WHITEHURST, Virginia
FLOYD D. SPENCE, South Carolina
DAVID C. TREEN, Louisiana
ROBIN L. BEARD, Tennessee
DONALD J. MITCHELL, New York
MARJORIE S. HOLT, Maryland
ROBERT W. DANIEL, JR., Virginia
ELWOOD H. (BUD) HILLIS, Indiana
DAVID F. EMERY, Maine
PAUL S. TRIBLE, JR., Virginia
ROBERT E. BADHAM, California
CHARLES DOUGHERTY, Pennsylvania
JAMES COURTER, New Jersey
MELVIN H. EVANS, Virgin Islands

JOHN J. FORD, *Staff Director*

FOREWORD

This report was prepared at the request of the Chairman, House Armed Services Committee. Its primary focus is on the history and current status of Vertical or Short Take-off and Landing (V/STOL) aircraft technology and is intended to assist Committee members in their review of Department of Defense V/STOL programs. The views contained herein are those of the author and do not necessarily reflect those of the Committee.

This report is primarily a review of the technology base supporting the development of Vertical or Short Takeoff and Landing (V/STOL) aircraft. However, since the concept of air-capable ships is intimately associated with V/STOL aircraft, it is most difficult to split the two and cover V/STOL only. Therefore, Section II* offers perspectives on the highly controversial subject of air-capable ships and their aircraft.

The review covers the period from the early 1950's to the present -- over a quarter of a century. During this time at least 55 V/STOL vehicles have been designed, built, and flight-tested. Obviously, it is not appropriate to cover all of these vehicles here. Rather, the various V/STOL vehicles were first classified according to their type of lift generation mechanism: jet lift; fan-in-wing/body; deflected slipstream; tilt wing; tilt prop/rotor; ducted prop; augmenter; and new concepts. A technology summary was then prepared for representative aircraft from each class, with due consideration to time frame and national origin. Twenty-one technology summaries are contained in Section III. Section III also highlights the successes, failures, problems encountered, and hopes for the future.

*In fairness, Section II should really be signed by Professor David C. Hazen of Princeton University. It contains, for the most part, his reflections on the problems confronting naval planners ("V/STOL and the Naval Planner's Dilemma," Astronautics and Aeronautics, June 1977).

(III)

Technology assessments are contained in Sections IV and V. Section IV discusses technologies having the most interactive effects on V/STOL aircraft. The methodology used to assess the state-of-the-technology is the predictive and measurement capability for each technology area. Section V points up specific problem areas and surfaces key issues in a much more succinct manner -- it is recommended reading for those with a less technical bent.

Since the Navy is faced with an impending commitment to V/STOL aircraft, some discussion of the Navy's envisionment is appropriate. Section VI provides this insight and discusses how the Navy plans to implement conversion to V/STOL aircraft.

Section VII provides a detailed funding history of V/STOL aircraft programs and supporting technologies since the early 50's. Only those programs funded by the Services are identified.

The <u>Executive Summary</u>, Section I, highlights the advantages and disadvantages of an air-capable Navy and lists the major problems yet to be resolved before this concept can become a reality. It is recommended for those not wishing to review the details shown in the supporting sections. Among the topics covered in the <u>Executive Summary</u> are the Navy's envisionment of an air-capable fleet, unresolved issues, overall technology base assessment, potential roles and missions of V/STOL aircraft in sea control, and major implications of shifting to V/STOL aircraft.

Much of the material in this study is drawn from, or based on, the extensive list of references exhibited in Section VIII. Not shown in these references, however, are the numerous private discussions conducted with industrial and DOD personnel associated with various aspects of V/STOL aircraft development -- both "yea" and "nay".

CONTENTS

	Page
List of Figures	VI
List of Tables	VII
List of Abbreviations and Acronyms	VII
I. Executive Summary	1
1. The Navy's Envisionment	1
2. Problem Areas	2
3. Summary of the V/STOL Technology Base	5
4. V/STOL Aircraft in Sea Control	9
5. Major Implications of Shifting to V/STOL Aircraft	20
II. V/STOL-Ship Perspectives	23
III. V/STOL Aircraft: Past and Present	31
1. Air Test Vehicle (ATV)	33
2. X-14/X-14A	34
3. SC.1	37
4. DO-31	40
5. XV-4B	43
6. VJ-101C	46
7. VAK-191B	49
8. AV-8A	52
9. AV-8B	55
10. XV-5A	58
11. X-22A	60
12. XV-3	63
13. XV-15	64
14. XC-142A	66
15. CL-84	69
16. VZ-3	72
17. XV-4A	75
18. XFV-12A	78
19. Rotor Systems Research Aircraft (RSRA)	80
20. Advancing Blade Concept (ABC)	82
21. X-Wing	84
IV. V/STOL Technology Assessment	86
Methodology	86
Propulsion and Propulsion Induced Effects	89
1. Inlet Performance	90
2. Bleed Effects	94
3. Engine Thrust Modulation	94
4. Propulsion Sizing	98
5. Environmental Effects	99
6. Lift Generator Performance	100
7. System Losses	107
8. Ground Environment	110
9. Hot Gas Ingestion	116
10. Induced Forces and Moments	121
Stability and Control	128
11. Equilibrium Trim	128
12. Static Stability	131
13. Control Power	133
14. Gust Sensitivity	135
15. Coupling	137
16. Dynamic Stability	140
17. Height Control	142

	Page
V. Problem Areas and Issues	145
V.1 Concepts, Configurations and Systems Integration	146
V.1.1 Systems Considerations	148
V.1.2 Aerodynamic Issues	148
V.1.3 Control, Stability and Flight Dynamics	149
V.1.4 Ground Effects	151
V.1.5 Propulsion Induced Effects	151
V.1.6 Aircraft-Ship Interface	152
V.1.7 Reliability and Maintainability	153
V.2 Propulsion System	154
V.2.1 Propulsion Technology Comparison	155
V.2.2 Reliability	155
V.2.3 Critical Propulsion Technologies	157
V.3 Vehicle Stability and Control	159
V.3.1 Control Characteristics	160
V.3.2 Vehicle Aerodynamics	162
V.3.3 External Environment	162
V.3.4 Stability Augmentation System	163
V.3.5 Other Important Issues of the Control System	163
V.4 Materials and Structures	164
V.4.1 Characteristics of Composite Materials	165
V.4.2 Failure Modes	167
V.4.3 Candidate Materials	168
V.4.4 Manufacturing Technology	168
V.4.5 Design Criteria	169
V.4.6 Composites in Propulsion Systems	169
V.5 Avionics	170
V.5.1 Flight Control	171
V.5.2 AEW Radar	172
VI. Review of Navy V/STOL Aircraft Plans	175
VII. V/STOL Funding Profiles	183
VIII. References	199

LIST OF FIGURES

	Page
I-1—V/STOL Vehicles	6
I-2—Summary—V/STOL Technology Evaluation	10
I-3—V/STOL Aircraft Summary	11
III-1—Air Test Vehicle (ATV)	33
III-2—X-14/X-14A	35
III-3—SC.1	38
III-4—DO-31	41
III-5—XV-4B	44
III-6—VJ-101C	47
III-7—VAK-191B	50
III-8—AV-8A	53
III-9—AV-8B Technology Improvements	56
III-10—AV-8B	57
III-11—XV-5A	59
III-12—X-22A	61
III-13—XV-3	63
III-14—XV-15	65
III-15—XC-142A	67
III-16—CL-84-1	70
III-17—VZ-3	73
III-18—XV-4A	76
III-19—XFV-12A	79
III-20—Rotor Systems Research Aircraft (RSRA)	81
III-21—Advancing Blade Concept (ABC) Helicopter	83
III-22—X-Wing	85
IV-8-1—Ground Proximity Propulsion Induced Effects	112
IV-8-2—Ground Footprint Symbology and Region Definition	113
IV-8-3—Velocities and Temperatures in the Jet Footprint of XJ 99 Lift Engine	114
IV-10-1—Lift Jet and Lift Fan Interference Forces in Transition	124
IV-10-2—Typical Lift Loss Effects on Aircraft	125
VI-1—Approximate IOC's for Replacement Aircraft	178
VI-2—V/STOL Decision Points	181

LIST OF TABLES

	Page
I-1—Predictive Capabilities of V/STOL Technologies	7–8
III—V/STOL Aircraft Selected For Review	32
III-1—Method of Flight Control of ATV	34
III-2—Method of Flight Control of X-14	36
III-3—Method of Flight Control of SC.1	39
III-4—Method of Flight Control of DO-31	42
III-5—Method of Flight Control of XV-4B	45
III-6—Method of Flight Control of VJ-101C	48
III-7—Method of Flight Control of VAK-191B	51
III-8—Method of Flight Control of P. 1127, Kestrel, and Harrier	54
III-9—Method of Flight Control of XV-5A	60
III-10—Method of Flight Control of X-22A	62
III-11—Method of Flight Control of XV-3	63
III-12—Method of Flight Control of XC-142A	68
III-13—Method of Flight Control of CL-84	71
III-14—Method of Flight Control of VZ-3	74
III-15—Method of Flight Control of XV-4A	77
V-4-1—Comparison of Graphite/Epoxy Composites with Metals	166
VII-1—U.S. Army V/STOL Programs	178
VII-2—U.S. Navy V/STOL Programs	181
VII-3—U.S. Air Force V/STOL Programs	184

LIST OF ABBREVIATIONS AND ACRONYMS

AAW—Anti-Air Warfare
ACS—Air Capable Ship
AEW—Airborne Early Warning
AGARD—Advisory Group for Aeronautics, Research and Development
ASCM—Anti-Ship Cruise Missile
ASUW—Anti-Surface Unit Warfare
ASW—Anti-Submarine Warfare
CAP—Combat Air Patrol
C & D—Cover and Deception
CG—center of gravity
COD—carrier on-board delivery
CP—center of pressure
CTOL—conventional takeoff and landing
DLI—deck launched interceptor
ECCM—electronic counter-counter-measures
ECM—electronic counter-measures
EGT—exhaust gas temperature
EAS—equivalent air speed
EMCON—emissions control
EMI—electromagnetic interference
EW—electronic warfare
FCS—flight control system
FIGV—fixed inlet guide vanes
FOD—foreign object damage
ft—feet
g—gravitational acceleration
HASC—House Armed Services Committee
HGI—hot gas ingestion
HUD—head-up display
HVU—high value unit
IAS—indicated air speed
IFR—instrument flight rules
IGE—in ground effect
in—inch
IR—infra-red
kt—knots
lb—pound
MA—marine assault
MFHBF—mean flight hours between failures
MPA—maritime patrol aircraft

mph—miles per hour
MTTR—mean time to repair
NADC—Naval Air Development Center
nm—nautical mile
OGE—out of ground effect
OTH/DCT—Over-The-Horizon Detection, Classification and Targeting
PIO—pilot-induced oscillations
QCSEE—Quiet, Clean, Short Haul, Experimental Engine
RCS—reaction control system
RF—radio frequency
ROA—radius of action
RPM—revolutions per minute
SAR—search and rescue
SAR—synthetic aperature radar
SAS—stability augmentation system
SFC—specific fuel consumption
shp—shaft horsepower
SOA—speed of advance
STO—short takeoff
STOL—short takeoff and landing
STOVL—short takeoff and vertical landing
TAS—true air speed
TIT—turbine inlet temperature
TOGW—takeoff gross weight
TOS—time on station
T/W—thrust-to-weight ratio
URG—underway replenishment group
VFR—visual flight rules
VIFCS—VTOL Integrated Flight Control System
VIGV—variable inlet guide vanes
VP—variable pitch
VSS—variable stability system
V/STOL—vertical or short takeoff and landing
VTO—vertical takeoff
VTOL—vertical takeoff and landing
WOD—wind over deck

I. EXECUTIVE SUMMARY

I.1 The Navy's Envisionment

An orderly modernization and evolution of sea-based aircraft requires a continuing reappraisal of the roles which can be fulfilled by aircraft in concert with new naval vessels and in new tactical concepts. Recognizing the advantages and utility of sea-based aircraft, the Navy maintains a continued effort to develop aircraft which could operate from not only carriers and amphibious support ships as do the fixed wing aircraft or helicopters today, but which, utilizing the maintenance support capability of these forementioned ships, could be staged or temporarily based from smaller ships. Aircraft operating from a broad spectrum of ships may enhance the individual mission effectiveness of those ships and increase overall force effectiveness. The most distinctive characteristic of such an air-capable Navy would be the wide availability of ships which could serve as bases for an ever-expanding variety of air vehicles, the most common of which would be helicopters and V/STOL airplanes. Ships which have been suggested for potential V/STOL operations are:

- SCS (Sea Control Ship)
- VSS (V/STOL Support Ship)
- CVV (V/STOL compatible carrier)
- LHA, LPH, LPD, LSD (amphibious assault ships)
- CSGN (nuclear powered strike cruiser)
- Spruance class destroyers (DD 963)
- Aegis-equipped DDG-47
- Fleet frigates (FFG-7 class)
- Replenishment ships
- Container ships
- SES (Surface Effect Ships)
- SWATH (Small Waterplane Twin Hull)

(1)

As the above shows, the envisionment of an air-capable Navy is one in which tactical aircraft are spread aboard a large number of small ships, rather than concentrated aboard a very small number of very large ships. Some of the advantages claimed for such a Navy are:

- widely dispersed forces
    - makes enemy's shore based weaponry less effective
    - compounds enemy submarine problems
    - confounds satellite reconnaissance
    - complicates enemy's efforts to mass forces
- more tactical aircraft at sea
- forces would be more readily available where and when needed
    - shorter reaction time
    - quick deployment to scene of action
    - more flexibility in assigning coverages
    - spot reconnaissance
- allows more efficient allocation of forces: permits small forces to match small tasks
- lowers risk of incurring out-of-proportion losses
- improved surveillance coverage
- smaller ships are more easily masked at sea and when leaving port
- V/STOL operations are not tied to carrier's cycle
- aircraft would be integrated closely with surface warship and its weapons
- more efficient use of surface ship's new weaponry
- ASW (Anti-Submarine Warfare)
    - sensor platform located far from ship's disturbances
    - decreases reaction time to attacks
    - aids in search, detection and localization of contact
- AEW (Airborne Early Warning)
    - aircraft serves to "raise-up" ship's antennas

I.2 Problem Areas

An audit of defense-related V/STOL research and development efforts since 1950 reveals an investment of over one-billion dollars. This estimate is very conservative since an accurate count of all programs (and dollars) stretching back almost 30 years cannot be obtained. In addition, the contributions of NASA, academia, foreign governments and company ventures are not included. Also, the costs associated with the Harrier are not included in the estimate. After almost thirty years and over one-billion dollars, the question is: What are the significant technological and operational "lessons learned" and what are the major problems which must be resolved before the concept of an air-capable Navy can become a reality?

Technological advances have been made which make V/STOL a potentially attractive weapon system concept for the 1990's and beyond. The combination of increased performance and weight reduction increases the possibility of developing an aircraft that may be competitive with conventional aircraft in performance and cost. In its present state, however, V/STOL carries significant extra development and logistics support costs as well as significant technological risk. In addition to assessing the technical feasibility of V/STOL designs, the cost/effectiveness of equally advanced conventional aircraft should be considered and compared to V/STOL designs.

Having highlighted the alleged advantages of an air-capable Navy, the following are some of the major technological and operational problems/ limitations which must be resolved.

- the largest obstacle: suitable, practical and financially feasible development of appropriate V/STOL aircraft
- limited space, weight and personnel accommodations in small ships
- selection of appropriate tradeoffs in building and arming ships
- ranges, payloads and operational experience with V/STOL are considerably less than CTOL aircraft
- on-board instrumentation severly limited for V/STOL: small and weak in comparison with those on-board CTOL aircraft
- V/STOL life expectancy in service use
  -engine long term reliability is degraded
  -more frequent checking of engines

- greater replacement or rework of engines
- shorter fuselage/wing service life
- how will command and control be implemented for greatly dispersed forces?
- aircraft-ship interface
  - limited visibility forward and downward for V/STOL
  - launch/recovery/approach procedures
  - when, where and how to transition to hover mode?
  - highly accurate, very close-range positioning information required from ship to pilot
  - smaller decks in bad weather and night operations
  - influence of sea-state: pitching and heaving deck
  - retrieval apparatus
  - tie down
  - winching the aircraft into tight hangar space
- safety

Other key areas that must be mastered before the idea of an air-capable Navy can come to fruition are:

- new propulsion technology must be developed
- miniaturized avionics must be developed to save weight
- must have significant increase in the use of composite materials to save weight
- ship and aircraft should be viewed as a total system, e.g.
  - what is the minimum size air-capable ship?
  - how far should aircraft and ship designs be compromised to be compatible?
  - maintenance and logistics concepts?
- aircraft weapons
  - small ships have small ordnance handling facilities
  - ultra-reliable, all-up munitions are required
  - more lethality per pound
  - more accurate
  - variety must be reduced
  - munitions must be standardized

I.3 Summary of the V/STOL Technology Base

Implicit in the Navy's commitment to a complete transition to V/STOL aircraft is operation from small ships under varying environmental conditions. Such operation poses severe requirements on several key technology areas, with the foremost among them being performance, handling qualities and propulsion systems. With regard to these key technologies, the question arises as to whether the existing technology base is adequate for design of V/STOL aircraft. Unfortunately, it is not — much of the technology for realizing this goal has evolved from a relatively sporadic development history stretching over three decades. Furthermore, with the possible exception of the Harrier, there is great concern over the lack of information regarding operational experience with V/STOL aircraft, particularly with respect to small platform operations at sea and the design guidance necessary for successful vehicle development. The planning of effective near-term and future R&D programs and formulation of vehicle requirements are seriously hampered by the uncertain state of V/STOL technology in general. Figure I-1 illustrates that, although many attempts have been made to produce viable operational V/STOL vehicles, only a few successful concepts have emerged to even operational prototype stages.

One measure of the status of a technology is its capability to adequately predict the characteristics of future vehicles. Therefore, by examining the state-of-the-art in predictive capabilities for key technology areas, it is possible to gain an overall assessment of today's V/STOL technology. See Table I-1. The predictive capabilities shown in Table I-1 are rated as follows:

Rating	Definition
None (N)	Little or no predictive capability for the actual vehicle characteristics; error (E) between the predictive and the final design is too large to provide even sound engineering estimates: $E \geq 50\%$

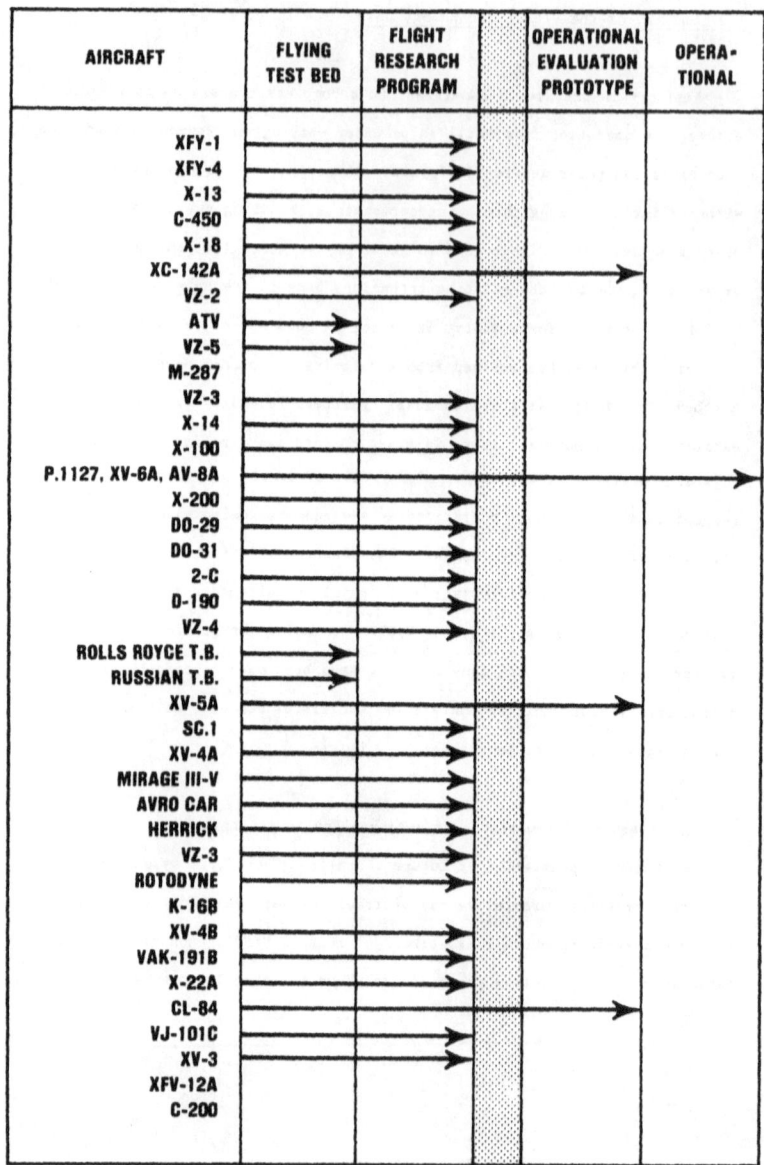

Fig. I-1. VSTOL Vehicles

Table I-1. Predictive Capabilities of VSTOL Technologies

TECHNOLOGY AREA / MEASURE OF TECHNOLOGY STATUS	THEORETICAL PREDICTION	MODEL PREDICTION	RIG PREDICTION
INLET PERFORMANCE	P	P	G
BLEED EFFECTS	N	P	P
ENGINE THRUST MODULATION	N	P	P
PROPULSION SIZING	G TO P	-	G
ENVIRONMENTAL EFFECTS	G TO P	-	G
LIFT GENERATOR PERFORMANCE	P TO N	P TO G	G TO P
SYSTEM LOSSES	P	P	G TO P
GROUND ENVIRONMENT	P TO N	P	P TO G
HOT GAS INGESTION	P TO N	P	P TO G
INDUCED FORCES & MOMENTS	P TO N	P	P TO G
EQUILIBRIUM TRIM			
IGE	N	N TO P	P
OGE	P TO N	P TO G	G
STATIC STABILITY			
IGE	N	N TO P	P
OGE	P TO N	P TO G	G
CONTROL POWER			
IGE	N TO P	P TO N	P
OGE	P	P TO G	G
GUST SENSITIVITY			
IGE	N	P	N
OGE	N TO P	G	N

Table I-1 (cont'd). Predictive Capabilities of VSTOL Technologies

TECHNOLOGY AREA / MEASURE OF TECHNOLOGY STATUS	THEORETICAL PREDICTION	MODEL PREDICTION	RIG PREDICTION
AERODYNAMIC CONTROLS			
IGE	N TO P	P TO N	.
OGE	P TO N	P TO G	.
CONTROLS BLENDING	.	.	G
ENGINE GYROSCOPIC EFFECTS	G	P	G
AIRCRAFT INERTIA	G	.	.
DYNAMIC STABILITY			
IGE	N TO P	P	.
OGE	P	G	.
HEIGHT CONTROL			
POWER			
IGE	P	.	.
OGE	G	.	.
DAMPING			
IGE	P	G	.
OGE	G	G	.

Poor (P) Predictive capability sufficient for initial estimates but inadequate to formulate detailed design characteristics:

$$10\% \leq E < 50\%$$

Good (G) Predictive capability adequate to define final vehicle characteristics:

$$E < 10\%$$

The first column in Table I-1 includes the applicable theory employed in the initial design, while the last two columns involve the predictive capabilities of data gathered from tests on similar or identical configurations (model on rig) that have been compared with flight test data. The following generic types of lift-generation systems were reviewed in establishing the rating for each technology area:

- Jet-Lift
- Fan-in-Wing/Body
- Ducted Prop
- Tilt Prop/Rotor
- Deflected Slipstream
- Augmenter

An average over all of the technology areas in Table I-1 yields the composite summary of the state-of-V/STOL-technology shown in Figure I-2. Table I-1 and Figure I-2 support the view that "the age of high performance V/STOL aircraft is that of infancy-not even adolescence." It appears that those individuals advocating "V/STOL now" are on shaky ground.

To compound the problem at hand, the V/STOL technology matrix is complicated by the burdensome variety of V/STOL aircraft which have been designed, developed and tested over the past thirty years or so. An appreciation for the large variety of approaches to V/STOL aircraft can be gleaned from the McDonnell Douglas V/STOL Aircraft Summary, Figure I-3. A common impression of V/STOL design technology is that it is configuration, rather than concept, oriented. Figure I-3 supports this view. The survey identifies fifty-five vehicles in several different aerodynamic and lift/engine configurations. These aircraft are spread over twenty-seven aircraft companies from seven countries — the United States, the Federal Republic of Germany, the United Kingdom, France, Italy, Canada, and the USSR. At the time of the survey, thirty aircraft were flying test beds (prototypes) or research vehicles; ten were under construction, in wind tunnel testing or mock-up; and fourteen were in the design concept phase. The AV-8A Harrier was, and still is, the only operational V/STOL aircraft in the Free World.

I.4 V/STOL Aircraft in Sea Control

The impact of V/STOL aircraft is separated into three categories:

1. Strategic and tactical benefits derived from the use of smaller air-capable platforms.

2. Operational benefits derived from the elimination of catapults and arresting gear.

3. Operational benefits derived from vertical launch and recovery operations.

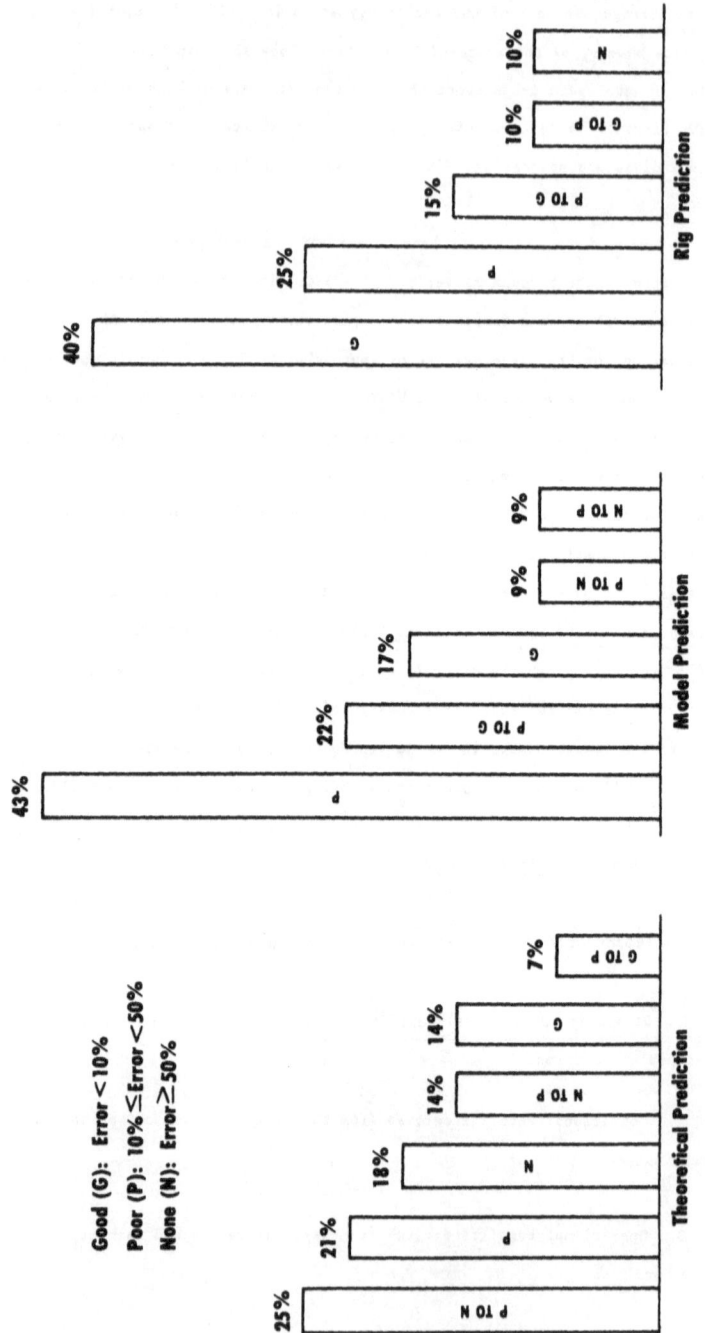

Fig. I-2. Summary-VSTOL Technology Evaluation

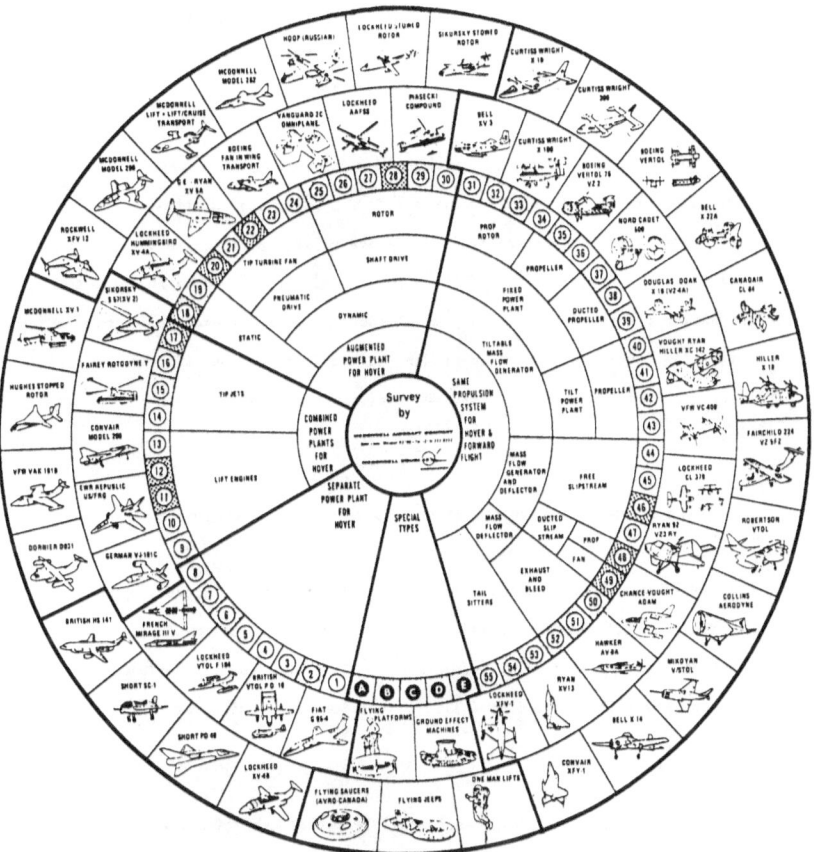

Fig. I-3. VSTOL Aircraft Summary

Benefits Derived From Smaller Air-Capable Ships

Reducing the size of an air-capable ship (ACS) produces two benefits. First, air assets can be dispersed more widely than is possible with a few large carriers. This follows from the fact that V/STOL's can be deployed on some existing non-carrier ships and from the assumption that smaller ACS's will be procured in large numbers. Secondly, a small ACS is more difficult to target than a large CV.

Dispersal of air assets is defined as the basing of aircraft on many platforms. The additional air-capable platforms may be used to provide greater geographical coverage, or to increase the number of ACS's in existing task groups, or both. Thus, the benefits of dispersing sea-based aircraft are contained within the concepts of <u>strategic dispersal</u> and <u>tactical dispersal</u> of aircraft.

Strategic Dispersal of Aircraft

Strategic dispersal is the simultaneous deployment of air assets to numerous geographical areas. V/STOL's potential for multiplying the number of ACS's can be converted into improved geographical coverage with sea-based aircraft. The result is air support in areas which otherwise would have no air cover due to the non-availability of a large CV.

The need for fixed-wing aircraft performance in future sea control missions has been clearly demonstrated. The advantages of fixed-wing aircraft over conventional helicopters are range, speed, and altitude capability. Range can be converted to offensive reach, speed to improved reaction time against threats, and altitude to longer sensor and communication ranges. Significant improvements in the sea control capability of a non-CV task group can be achieved with the addition of a small (i.e., V/STOL) air group.

The greater geographical coverage inherent in strategic dispersal of aircraft would be of value in at least two scenarios. In a major war at sea, small V/STOL air groups could provide air cover for task groups, such as convoy and URG (Underway Replenishment Group) escort, which would probably not have the support of a large carrier. In peacetime, small V/STOL air groups would improve the visibility, reaction time, and combat credibility of dispersed presence forces.

A corollary to strategic dispersal is the concept of air group tailoring. Tailoring is integral to the task force concept. With the present CTOL-only force, the Navy does not have the ability to tailor air group size to the threat. By contrast, the size of a V/STOL air group could be tailored for the mission and expected threat level by including the appropriate number of small ACS's.

Tactical Dispersal of Aircraft

Tactical dispersal is the dispersal of air assets among several ships within a task group. Two implications of tactical dispersal are significant:

1. The effect of disperal on air group survivability.

2. The opportunity to deploy dispersal air assets to the perimeter of an open formation.

Air Group Survivability

The loss of air capability sustained in an anti-ship cruise missile (ASCM) attack is a function of the following factors, et al:

- probability that an ASCM will select an ACS as its target.

- probability of ASCM kill by area and point defense systems.

- cost of a ASCM hit, in terms of air capability. The loss includes damage to the flight deck and critical equipment as well as aircraft destroyed.

Aircraft dispersal has an effect on all of these factors and the interrelationship is complex and not well understood. For example, the point defense capability of each ACS would be expected to increase with ship size; however, the effectivenss of area defenses and overall saturation level of the task group might be increased by greater dispersal. Furthermore, it is obvious that increasing dispersal will clearly lower the cost of a ASCM hit, since a single missile hit cannot damage the air capability of more than one ship. On the other hand, a large ship is less likely to suffer crippling damage to critical flight operations equipment than a small ship. Additional studies are required to clarify the effect of aircraft dispersal on ACS defensive capabilities and the amount of air capability lost from a missile hit.

Whether the aircraft are dispersed or concentrated, survivability is greatly enhanced by basing the air group on ACS(s) which cannot be easily targeted. Size is critical to targeting denial. The most that can be said at this point is that dispersal probably would increase the chance that some air capability would be lost in a missile attack but would reduce the chance of a total loss of capability.

Open Formation and Perimeter Dispersal of V/STOL

Tactical dispersal of V/STOL's presents an opportunity to deploy ACS's on the perimeter of an open formation. An open formation is a formation in which some or all of the ships are separated from each other by ranges which preclude conventional means of mutual support. In theory V/STOL's would provide mutual support in such a formation. Relative to close formations, the benefits of perimeter dispersal of V/STOL's are increased radius-of-action (ROA)/time-on-station (TOS) for the aircraft and quicker response to threats approaching the perimeter. Although the quicker response also can be provided by perimeter air loiter of aircraft (e.g., CAP (combat air patrol)) flying from the center of the formation, an open formation would permit perimeter deck loiter. This would reduce the number of aircraft required to keep one on station.

Despite these benefits, open formations suffer from several drawbacks which make them of questionable value. The following weaknesses impact directly on V/STOL effectiveness:

- Because task group assets cannot provide the threat warning required to employ dispersed V/STOL's in a reactive mode, the pickets must be supported by an external surveillance system. To provide AEW on the perimeter requires each picket to carry several AEW aircraft (so that one can be kept in the air), or a large detachment of AEW aircraft based in the formation center. Detection of sub-threats would require either the use of air-dropped sensors around the perimeter (again requiring many aircraft) or the use of many pickets to provide overlapping sonar coverage.

- Although V/STOL's provide mutual support between ships in ASW (Anti-Submarine Warfare) and ASUW (Anti-Surface Unit Warfare), mutual AAW (Anti-Air Warfare) support is lost due to the excessive time required for aircraft to concentrate at any point on the perimeter.

The conditions required for open formation/perimeter dispersal of V/STOL's to be effective are an external surveillance system and a minimal threat from aircraft or long-range ASCMs.

Reduced Targeting Vulnerability of Air Capable Ships

The smaller size of V/STOL platforms would reduce the ability of the enemy to target them using sensors such as radar and IR which rely on size discrimination. Reducing the targeting vulnerability of ACS's is crucial for the following reasons:

1. The capabilities of sea-based, fixed-wing aircraft render any ACS a high value target.

2. It will be very difficult to hide a surface task group for prolonged period from future surveillance systems. Detection of the force should be assumed and emphasis shifted to preventing the enemy from localizing/ classifying the HVU's (High Value Unit).

3. Concealing a large carrier within a group of much smaller vessels is a formidable challenge for cover and deception techniques. Conversely, it should not be difficult to conceal a V/STOL ship in a merchant convoy, URG, or surface task group by use of EMCON (Emissions Control-limited use of active radars, communications, data links and other RF radiators).

4. Preventing the launch of future extended-range ASCM's may not be practical in many cases. This lends greater urgency to preventing classification of the HVU's. Without discrete targeting of the HVU's, the enemy is forced either to expend a much greater number of missiles to achieve the same kill probability against desired targets, or to resort to high-risk tactics by manned reconnaissance aircraft to obtain a visual classification.

In short, permitting a reduction in the size of the ACS such that its identity can be masked from enemy sensors may prove to be one of the greatest tactical benefits of V/STOL. Additional study is required to determine how small ACS's must be to achieve this benefit and the importance of size discrimination in future enemy targeting techniques.

Benefits Derived From Reduced Reliance on Catapults and Arresting Gear

More Reliable Launch and Recovery Operations

Present aircraft carrier operations are sometimes degraded by technical difficulties with catapults and arresting gear. This problem would be magnified on smaller CTOL carriers with less redundancy of equipment. V/STOL aircraft could eliminate such problems as a source of launch and recovery delays. The tactical benefits would be more rapid and reliable response by alert aircraft and elimination of aircraft diverts/losses due to arresting gear failures.

Reduced Susceptibility to Flight Deck Damage

Due to the absence of special launch and recovery equipment, the air capability of a V/STOL ship should be more resistant to damage than that of a CTOL ship of comparable size. However, if the V/STOL ship were smaller, this effect could be outweighed by the overall inferior resistence to damage of a smaller ship.

Elimination of Catapult Noise

The elimination of catapults would deprive an enemy of catapult noise which is presently a primary means of long-range CV classification and coarse location.

Benefits Derived From Vertical Takeoff and/or Landing

Elimination of Wind-Over-Deck Requirements

Vertical flight operations should be practical in most crosswind conditions, thereby eliminating the need to maneuver the ACS into the wind. The result would be a reduction in the targeting vulnerability of the ACS and an increase in speed of advance.

Effect on ACS Targeting Vulnerability

The need to turn into the wind periodically is a characteristic of aircraft carriers which can be exploited by the enemy to identify the carrier within a formation of ships. Many considerations, such as formation size, EMCON, and the speed for escorted ships, limit the practicality of turning the entire formation into the wind for every launch and recovery evolution. C&D (cover and deception) techniques can be used to prevent the enemy from determining exactly which ship is the carrier. However, prolonged observation of the force by satellites or MPA's (Maritime Patrol Aircraft) should succeeed in eliminating some potential targets and thus increase the probability of a missile hit on the carrier. By contrast, an ACS whose aircraft are operating in the VTOL mode would not be vulnerable to this method of targeting.

Increase Speed of Advance

Eliminating WOD (wind-over-deck) requirements also increases the SOA (speed of advance) of air capable ships. Sea control exercises with large CV's have demonstrated the difficulty of maintaining even a moderate SOA during "flex-deck" operations, which require frequent launches and recoveries. If all launches and recoveries were vertical, a higher SOA could be maintained.

Reduction in ACS Radiated Noise

Since the ACS does not have to steam at high speed to generate the WOD and to regain station after flight operations, acoustic radiation is reduced with consequent reduction in ACS vulnerability to targeting.

More Expeditions Launches and Recoveries

Vertical launches and recoveries can be accomplished more expeditiously than CTOL and STOL operations because several aircraft can be launched and/or landed simultaneously. The result is improved response time for quick reaction aircraft such as DLI's (deck-launched interceptor), and faster turn-around times for returning aircraft.

Comparison of VTO and STO Operations

VTO Operations

Vertical flight operations maximize all of the above benefits at the expense of a significant degradation in performance. In the VTOL mode the V/STOL's could be flown from ships as small as a DD963, providing great potential for dispersal and complicating the enemy's targeting problem. VTOL operations impose a significant performance penalty relative to STO operations. An aircraft with improved VTO performance could be built, but it would probably be too heavy to operate from smaller platforms. The tactical implications of VTO performance are:

1. The performance penalty will be most significant in close formation tactics because in this case the offensive reach of the task group will be provided entirely by the V/STOL's radius of action. In theory, perimeter dispersal could compensate for the limited VTO radius of action. However, war gaming results suggest that perimeter dispersal is not a viable tactic in many situations.

2. In AAW (anti-air warfare), VTO performance will be adequate for DLI missions. Tasks requiring air loiter (CAP, AEW, EW) can be performed by VTOL's but time on station will be reduced. This will produce a higher backup ratio to keep one aircraft on station and creates additional opportunities for mission abort.

3. In ASW (anti-submarine warfare), prosecution range will be reduced with VTO operations. Reduced VTO range should provide adequate defense against torpedo and medium-range ASCM platforms, but eliminates any chance of destroying the long-range ASCM platforms.

4. In ASUW (anti-surface unit warfare), VTO range will be adequate for support of ASCM strikes. The reduced performance would be significant in the absence of satellite surveillance, since V/STOL's would then be required to maintain continuous surveillance of large ocean areas.

STO Operations

The use of STO launch techniques permits improved aircraft performance at the expense of an increase in the size of the ACS and loss of the benefits provided by vertical launches. The deck length required for STO launch varies considerably between aircraft designs, but 300-400 ft. is common. The "ski-jump" concept or low energy catapults could reduce this length. Nevertheless, it is a reasonable assumption that STO operations will require a larger flight deck than VTO operations. Thus, STO will not be compatible with as many existing platforms at the VTO mode, and ACS's designed for STO will be somewhat larger than VTO-only ships need be. However, it is probable that concealment can be accomplished with platform sizes large enough to enable STO operations.

Although STO Operations will normally require a turn into the wind, time spent into the wind can be minimized by limiting the number of aircraft per launch and accomplishing vertical landings when returning to station. These measures, with the use of a random formation and other C&D techniques, should limit the vulnerability of the ACS to targeting by observation of flight operations maneuvering.

The final disadvantage of STO mode is that aircraft cannot be launched as expeditiously as in vertical operations. A turn into the wind is required and only one or two aircraft can be launched simultaneously. These delays may dictate using the VTO mode for quick-reaction aircraft missions.

Tactical Implications of V/STOL Relative to CTOL

On a plane-for-plane basis, V/STOL's probably will be inferior in performance to the next generation of CTOL aircraft that could be built. If performance is held constant, the V/STOL will be heavier than a CTOL of comparable technology. This follows from the fact that an engine capable of sustaining vertical flight is oversized for most other mission requirements. Alternatively, the CTOL would have superior performance to a V/STOL of equal weight. This plane-for-plane weight/performance penalty is greater for the

Type A V/STOL (subsonic) than for the Type B (supersonic). The high thrust-to-weight ratio required for vertical flight is not required for subsonic cruise and loiter, which typify most Type A missions, but high thrust-to-weight is desirable for many Type B missions.

Since V/STOL will be constrained by the desire to operate from small ships, a performance rather than a weight penalty is more likely to be realized. Assuming aircraft of equal weight, the plane-for-plane advantage of CTOL over V/STOL would be superior range, payload and endurance. The speed and service ceiling of the two designs should be comparable.

Despite this performance penalty, V/STOL capabilities should be adequate for all sea control tasks except destruction of air and subsurface long-range ASCM platforms. It is not at all clear, in fact, whether advanced CTOL aircraft would substantially improve air group effectiveness against long-range threats.

V/STOL's plane-for-plane performance penalty relative to CTOL must be weighed against the strategic and tactical benefits of V/STOL discussed earlier. The proper criterion for assessing the relative utility of V/STOL and CTOL is not individual aircraft performance, but the contribution to overall fleet effectivness. Key issues remaining to be resolved are how much dispersal of aircraft is desired, and how small an ACS must be to prevent discrete targeting.

I.5 Major Implications of Shifting to V/STOL Aircraft

There is every reason to presume a continuing scarcity of Navy resources and a progressive increase in the foreign military threat. These facts underscore the necessity for increased effectivenss of operational forces if basic naval obligations are to be realized at year 2000. A shift to V/STOL aircraft would have far reaching implications and substantial risks. The weapons, the ships, the systems, the aircraft, the personnel — in short, practically the whole Navy — will be influenced in large degrees by the idea of an air-capable Navy. Indeed, a recent independent study group (chartered by the Navy) concluded that the V/STOL program is considerably more than the development of another new aircraft. It involves the advancement of many demanding technologies, a balancing of these technologies within the evolving aircraft program, and an extremely critical "give and take" interaction between the aircraft technology programs and programs to define and develop the mission requirements and operating concepts, the new platform basing options and logistics, the aircraft-platform interface(s), new sensors and most likely new tactics and weapon systems. All of these factors impact each other. A procedure is required to meet the demands of flexibility, interfaces, and the systems engineering/integration task which is virtually unprecedented by even such programs as Polaris, Poseidon and Trident because it impinges on such a large part of the Navy. Maintaining interface flexibility requires coordination between almost every Navy Command in order to effectively pursue and combine the current and future operational needs of the Navy and their translation to technical requirements — a major reconfiguration of the Naval surface/air power and operations is involved. This being the case, and recognizing the many "navies" within the Navy, one wonders how such integration/coordination can be forged. Perhaps, a V/STOL "czar" would be required?

II. VSTOL-SHIP PERSPECTIVES

The need for aircraft at sea has rarely been doubted. Their employment, however, has been paced by technological and operational progress, primarily in the aircraft, but also in the ships. The aircraft carrier has served as a reference for conventional aircraft for almost 50 years; and probably no other type of naval vessel in the course of history has provided the flexibility and power as is inherent in the carrier. By changing its complement of aircraft and mix of weapons, the carrier can be adapted to a large number of different missions. Because of its size, it can operate in sea states that force lesser ships to lose effectiveness. Size, protective armor, and designed redundancies permit it to absorb substantial damage. Accompanied by an appropriate screen and augmenting the defense by its own aircraft, the carrier is a formidable weapon system indeed.

It is also very expensive, not just to build, but also to operate and keep supplied with the skilled personnel necessary to make its sophisticated systems work. As a consequence, the Navy can afford only a limited number. As the size and complexity of carriers have increased, the number which can be afforded has decreased, dropping from 24 to 12 in the last decade. This fact alone has greatly increased the vulnerability of these ships by reducing the number of targets upon which the enemy must concentrate. Their small numbers coupled with their vast size, which makes them readily identifiable, makes them the object of a great deal of unwanted attention in the event of a major war. In view of the continuing proliferation of sophisticated weaponry, many strategists believe these ships are becoming unacceptably tempting targets in even less than major wars.

The realization that there is dwindling enthusiasm to pay the roughly two billion dollars for Nimitz - class carriers has caused the Navy to start thinking in terms of smaller carriers and more air-capable ships . . . a Navy in which tactical aircraft are spread aboard a large number of small ships, rather than concentrated aboard a very small number of very large ships. This in turn has caused a quiver of anticipation to run through the VSTOL community. Maybe because of the Soviet's VSTOL carrier Kiev, smaller carriers seem to many to signal the arrival of a clear cut operational requirement for VSTOL. However, the U.S. Navy's requirements are far different from the Soviets' and the desirability of such ships and planes in fulfilling its missions is still highly debatable.

Soviet Naval Planning

At the end of WWII, the U.S. emerged with the world's most powerful fleet and with the world's most devastating weapon. It is not surprising that the Soviet Union perceived a major threat arising from the combination of the two, and set out to design a force to neutralize it. This effort underwent a number of false starts owing to the long ingrained land force traditions of the country, but eventually took the form of a balanced surface, subsurface and air force equipped with a variety of cruise missiles specifically designed to strike at U.S. carrier task forces.

A major effort of such magnitude requires time to come to fruition and not until the 60's did the ships, planes and weapons of the Soviet fleet emerge as a coordinated anti-carrier system. By that time the threat was already changing. The intercontinental ballistic missile had arrived and gone below the waves. Where once the highly mobile carrier forces had represented the major problem confronting Soviet naval planners, neutralizing the missile-carrying nuclear submarines became their objective.

Like the U.S. Navy, the Russians view the submarine as the primary vehicle for conducting operations against other submarines. This being the case, it follows that not only should the Soviets try very hard to develop improved methods of underwater surveillance in order to detect, track and target U.S. submarines, they must also work very hard to prevent us from operating in a similar manner against their submarines.

There are strong indications that the Soviet navy is building with such objectives in mind. Their earlier surface ship designs were capable of very high speeds and equipped with substantial firepower; but they had relatively short range and only limited reload ability. Such ships could make devastating pre-emptive attacks on carrier groups closing the coast to launch a strike, but were relatively ineffective for any ocean control role.

More recent construction, while retaining the heavy firepower, is beginning to emphasize range and endurance. The "extended coastal defense" concept is giving way to that of the world-wide navy, less with the idea of projecting power ashore — long a motivating force behind U.S. naval development — then with the idea of providing anti-ship support to their submarines. Viewed in this context, the Kiev, with its heavy anti-ship armament coupled with VSTOL fighters makes a lot of sense. It is not the Soviet navy's intent to engage in carrier group versus carrier group warfare, in the style of WWII. That's what long-range anti-ship missiles are for. Rather, the Kiev-class seems designed to deny airborne antisubmarine warfare forces the immunity they have enjoyed when out of range of of shore-based fighters. A Yak-36 "Forger" may not be much of a challenge for an F-14, but it could give a P-3 fits.

Even though the Soviet navy now has only two helicopter carriers (the Moskva and Leningrad) and the new Kiev, naval aviation plays an important role. Aircraft such as the "Badger" with a tactical range of 1500-2000 mi. and "Backfire", with even greater range, carry anti-ship missiles of the 100-150 mile class. Additionally, these aircraft have the ability to take over the guidance of ship-or submarine-launched missiles. And they can be refueled in flight to further extend their range. The "Bear-D" and specifically configured "Badgers" are used as long-range reconnaissance aircraft to augment the satellites used for this purpose — currently of limited, but growing, capability.

The Soviet navy's development will most likely continue along the line already established. Ocean surveillance by satellite will continue to improve, as will the already efficient command, control and communication system. ASW procedures, sensors and equipment will receive continuing developmental effort as will surface, subsurface and air-launched anti-ship missiles carrying both conventional and nuclear warheads. More Kiev-class vessels will be built. It is not clear whether, as they grow more familiar with carrier operations, the Russians will become more ambitious with respect to size in order to reap the benefits of additional fuel capacity and larger aircraft complements. Even so, they will probably refine their VSTOL efforts to produce a supersonic fighter more competitive with U.S. carrier aircraft.

U.S. Naval Planning

In light of the Soviet naval build-up, the U.S. Navy has been trying to adapt its forces within the limits of budgeted funds to meet its assigned missions. The order of priority and the detailed nature of the Navy's missions are being debated; but surely the nation expects the Navy to provide it with defense against enemy forces and to be a visible and tangible instrument of national policy, capable of projecting U.S. force around the globe. Unlike the Soviet Union, we depend greatly upon foreign trade, the vast majority carried by water; and most of our allies are overseas. Control of the seas is much more important to the U.S. than to the USSR. As both the Russian overseas influence and merchant fleet grow, this difference will lessen.

However, this primary difference governs the decisions about the composition of the two naval forces. The Soviets have but one potential threat and therefore can concentrate their efforts on seeking means of defeating the U.S. Navy and denying this country the use of the seas, while our Navy must be prepared to tailor its activities from a show of force, to crisis control, to limited conflict, to all-out conventional war, to nuclear holocaust. While the Soviets are thus free to produce forces of a specialized nature, flexibility is the controlling factor for the U.S.

Unfortunately, the units making up naval forces are not infinitely adjustable. Even relatively small vessels require several years to move from concept to an operating entity; and an aircraft carrier may take seven or more years to complete. The naval planner must thus try to achieve the best match he can between projected capability and perceived threat by building as much flexibility as possible into a relatively inflexible number of platforms.

In an environment in which its most effective tactical weapon system--the carrier and its strike aviation complement--are becoming more vulnerable both because of growing Soviet capabilities and an economically imposed force-level limitation, the Navy must plan its development programs to meet that needed flexibility. In the recent past, the effort seems to have been directed primarily at trying to convince Congress of the necessity of building more carriers. Even if this effort eventually pays off, the grudging manner in which the needed funds and authorization will be obtained must surely signal to even the most avid supercarrier enthusiast that the Navy needs an alternate solution.

VSTOL Potentials and Problems

Many such solutions are under study. One put forth a number of years ago was the Sea Control Ship — a relatively small (in comparison to the supercarrier) vessel of about 14,000 tons carrying a mixed complement of helicopters and hypothetical VSTOLs. It would carry out escort work by combining the helicopter's ASW ability with the VSTOL's air defense. Another potential solution, the VSTOL

Support Ship (VSS), a vessel of somewhat elastic tonnage, would carry hypothetical VSTOLs and operate, admittedly with lessened effectiveness, in much the same manner as the current carriers. Other possibilities looked at include smaller but nonetheless conventional, carriers employing conventional aircraft.

All studies show the most cost-effective solution to be the largest ship one can build. Ton for ton, a large ship can be built, equipped and manned more cheaply than two ships of half its size. The larger the ship, in general, the less sensitive it is to sea state, thereby providing more nearly all-weather capabilities, and the larger can be its aircraft complement and associated support. Perhaps even more importantly in this age of rapidly escalating personnel costs, the larger ship permits more effective use of a crew.

Such studies have also tended to produce less than wild enthusiasm for VSTOL aircraft. Contrary to the statements of some officials who would like to believe that the work in the field over the past three decades has placed on the shelf a canned technology that can be taken out and somehow quickly painted onto a new aircraft, development of a suitable machine is going to involve both costly and lengthy airframe and engine development as well as significant advances in avionics and flight handling qualities. Other key technological and operational issues that must be resolved with respect to an air-capable Navy are the air-ship interface problems, command and control, and logistical support.

Since to warrant the name, VSTOL aircraft must be able to do just that, i.e., takeoff and land vertically or with only comparatively short ground runs, most of the past development has been directed toward solutions to these aspects of the performance problem. Thanks to the lengthy development of the Pegasus engine and the airframe wrapped around it, the AV-8A Harrier has marginally acceptable performance for some applications. Certainly in the hands of the Marines it has performed remarkably well, although it has not, of course, been tested in actual combat. The advanced Harrier (the AV-8B), gradually moving towards existence, will have substantially better performance.

With the arrival of the AV-8B, the Marines will have a light attack aircraft able to go ashore with them and provide the type of close-air-support that they desperately require. It will operate primarily in the STOL mode, with VTOL an option if operational exigencies so demand. For this type of service it will have acceptable range/payload characteristics. But unless more is done, its handling characteristics, like those of the AV-8A, will leave pilots dissatisfied, and make it unsuitable for regular operation from a pitching, heaving deck under marginal weather conditions.

There is no particular reason why a naval version of the AV-8B cannot be produced. Indeed, the British plan one (the Sea Harrier), in which fly-by-wire techniques coupled with increased control power will provide suitable handling qualities. However, when equipped for naval strike missions with appropriate sensors and fire control systems on board, even when operated in the STOL-mode with substantial wind-over-the deck, it will at best be a light attack aircraft with a severely limited payload/range capability compared to existing carrier aircraft that can utilize in-flight refueling (presumably denied the Harrier pending the development of VSTOL tankers).

Even should the Harrier go to sea, it is a subsonic airplane. Although on sea trials it on several occasions demonstrated an ability to intercept Bears, it would be incapable of reaching a Backfire either before or after it had launched its missiles. The configuration of the Pegasus is such that its lumps and bumps come in awkward spots, making the design of a supersonic aircraft utilizing such an engine concept difficult.

In short, should the Navy set out to design a VSTOL aircraft with the capabilities of an F-18, for example, it would be confronted with the cost and time required for the normal development of such a high performance machine plus whatever would be needed to develop the VSTOL aspects. That could well involve the development of at least one, and probably more, new engines. The technology developed over the past three decades has shown that a supersonic VSTOL fighter is at least feasible; but to get it to the stage of operational reality is going to be a lengthly and expensive process — and one by no means devoid of technical risk.

The history of the development of VSTOL aircraft can quite aptly be described as a technology in search of an application. With the exception of helicopters and the Harrier, the search has not been very successful. Regardless of how obvious the advantages of packaging high performance within an airframe capable of short or vertical takeoff and landing may seem to VSTOL advocates, the people writing the checks to pay for aircraft development have remained unconvinced.

In spite of these rather gloomy forecasts, the Navy is undertaking an intensified VSTOL development program, with three types of machines under consideration, titled Type A, B, and C. Type A would provide a replacement for the Marine Corps assault helicopters and serve as an ASW and airborne early warning (AEW) platform. Type B would be a strike fighter. Type C would eventually replace the LAMPS (Light Airborne Multi-Purpose System) Mk III helicopter.

Type A will probably receive the most immediate effort because this subsonic machine involves less development risk. Normally LHA and LPH ships would carry Type A aircraft, but they should also be able to operate from all amphibious ships equipped with helicopter platforms. These aircraft may be equipped for ASW and/or AEW functions and flown from the larger classes of non-aviation ships. They would not be ready for the fleet until the mid-1980s at the earliest.

Type B is a much fuzzier and uncertain project. Given the current F-14 and F-18 commitments, the earliest such aircraft could be introduced into the fleet would be the mid-1990s — some twenty years hence. Type C is also at least twenty or more years in the future.

With the exception of the Type A, such IOC (Initial Operational Capability) dates will not stir the industrial pulse. Lacking the stimulus of a greatly accelerated arms race — which might lead to the construction of more supercarriers, thus putting off VSTOL development still further — nothing much earlier can be regarded as probable.

Options in the Face of Reality

Confronted by these realities the Navy must choose its strategy with care. The challenge is real and grows steadily as new Soviet ships are launched and additional Soviet aircraft join the inventory. Simultaneously, our carriers grow older. In twenty years ten of the current ships will be over 30 years old, and long before then we will either have to have started replacing them or to have found some viable alternative unless we wish to face a drastic reduction of our airpower over the seas. Given our reliance on the sea lanes, in face of developments such as Backfire and the Kiev, to lower the fleet's airpower would seem tantamount to a national death wish. Indeed, our current capacity to conduct air warfare over the oceans must not only be maintained but substantially augmented.

Alternative methods of achieving our goals must be carefully weighed. The U.S. Navy cannot afford to pursue many expensive programs, nor can the Navy afford to gamble on untried systems without providing adequate fall-backs. Inevitably, the fleet of the 1990s will be a blend of systems and abilities that in combination attempts to meet the requirements.

Whether or not VSTOL aircraft will constitute one of the elements in this blend will depend upon the result of development efforts over the next twenty years. At the moment, no VSTOL has been produced that approaches the capability of existing carrier aircraft. It may be that one need not think in terms of matching capabilities on an aircraft-to-aircraft basis; but the total system involving the aircraft, its weapons and its sea-borne base must produce a response that counters the threat projected for the period of its anticipated operational life. The ability of VSTOLs to provide adequate endurance without airborne refueling, or how to provide such refueling capability if it is required, is of particular concern.

Although VSTOL, when appropriately developed, may provide a greatly to be desired flexibility of basing, allowing a wide variety of ship platforms to carry high performance aircraft, only the foolhardy would gamble on replacing ships capable of launching and recovering conventional aircraft with vessels that can only handle VSTOLs when no such suitable aircraft yet exist. To hedge its bets, the next class of aircraft carriers the Navy builds will probably be smaller and less capable than the Nimitz class, but it will be equipped with catapults and arresting gear capable of handling the existing inventory of aircraft.

Long-range development plans should continue to explore VSTOL possibilities. Naval architects will be challenged to produce smaller, cheaper carriers with improved sea-keeping characteristics. Trade-offs between catapult size, ship speed and VSTOL characteristics (achieved in various ways) should receive constant study, as should alternative methods of performing anti-air and sea-control missions, such as long-endurance aircraft.

The imperative is the avoidance of the very big war. Although there are differing views of the roles and effectiveness of naval units, there is general agreement that, along with a credible nuclear deterrence, a clear capability of being able to acquit ourselves well in a big conventional war is essential to this avoidance. To provide this we must make sure that our side has the most effective weapon systems that we can invent. Equally, it would be nice to disperse our forces and reduce the size and value of individual targets. Smaller carriers — and, in general, the concept of an air-capable Navy — would permit the U.S. to move in this direction. The development of VSTOL aircraft coupled with appropriate weapons and support ships would allow carrying this dispersal still further. Thus one may entertain the feeling that naval VSTOL will come to pass, but its time has not yet arrived.

III. VSTOL AIPCRAFT: PAST AND PRESENT

This section provides technology profiles for twenty-one VSTOL aircraft representing the following eight basic categories of lift-generation:

- Jet-lift
- Fan-in-wing/body
- Ducted prop
- Tilt prop/rotor
- Tilt wing
- Deflected slipstream
- Augmenter
- New concepts

The aircraft selected for review are shown in Table III. In making these selections, primary consideration was given to the type of lift-generation, national origin and time period in which the aircraft were developed. Where appropriate, significant problems experienced during the development and testing of each aircraft are highlighted. The "special types" of VSTOL vehicles shown in the McDonnell Douglas survey (discussed earlier), e.g., flying platforms, flying saucers, ground effect machines, are not included in this review. The Lockheed XFV-1, Ryan XV13 and Convair XFY-1 are also excluded as "tail sitter" type aircraft are a wide departure from the principal VSTOL concepts being currently considered.

TABLE III. VSTOL AIRCRAFT SELECTED FOR REVIEW

Aircraft	Type	Time Frame	Who
ATV	jet-lift	mid 50's	Bell/U.S.
X-14	jet-lift	mid-late 50's	Bell/U.S.
SC.1	jet-lift	late 50's	Short/U.K.
DO-31	jet-lift	1960's	Dornier/Germany
XV-4B	jet-lift	mid 60's	Lockheed/U.S.
VJ-101C	jet-lift	mid 60's	MBB/Germany
VAK-191B	jet-lift	early 70's	VFW/Germany
AV-8A	jet-lift	1958-present	Hawker/U.K.
AV-8B	jet-lift	1973-present	McDonnell Douglas/U.S.
XV-5A	fan-in-wing	mid 60's	Ryan/U.S.
X-22A	ducted prop	mid 60's - present	Bell/U.S.
XV-3	tilt-rotor	50's-early 60's	Bell/U.S.
XV-15	tilt-rotor	early 70's-present	Bell/U.S.
XC-142A	tilt-wing	mid 60's	LTV/U.S.
CL-84	tilt-wing	mid 60's	Canadair/Canada
VZ-3	deflected slipstream	1950's	Ryan/U.S.
XV-4A	augmenter	early 60's	Lockheed/U.S.
XFV-12A	augmenter	present	Rockwell/U.S.
RSRA	new concept	present	Sikorsky/U.S.
ABC	new concept	present	Sikorsky/U.S.
X-Wing	new concept	present	Lockheed/U.S.

Identification: Air Test Vehicle (ATV)/jet-lift

Sponsor/Developer/Time Frame: Bell Aircraft Co./mid 50's

General Information: ATV was a high-wing direct-jet-lift research vehicle built as a company venture; test bed to substantiate conclusion of 1953 Air Force study that jet-powered VSTOL aircraft was feasible; first hover flight was in September 1954; it was the first jet VSTOL flown in the U.S.

Weight: 1000 lb.

Engine Characteristics: Two swiveling J44 missile engines mounted on a common shaft passing thru fuselage; rotation of shaft accomplished by hydraulic pressure for three complete rotation cycles.

Flight Control System: RCS attitude control for low-speed flight; four constant bleed flow nozzles: two located at rearmost point of aircraft (one provided pitch control and the other yaw control), remaining two located at each wing tip and operated differentially to provide a rolling couple.

Problems Encountered: Ground effects (heating of fuselage between engines); no complete transitions were performed; hover at altitude not possible (low static thrust); fountain effects; large coupling between pitch and roll because of gyroscopic effect of the vertical alignment of both engine spin axes.

Development Status Attained: Flying test bed

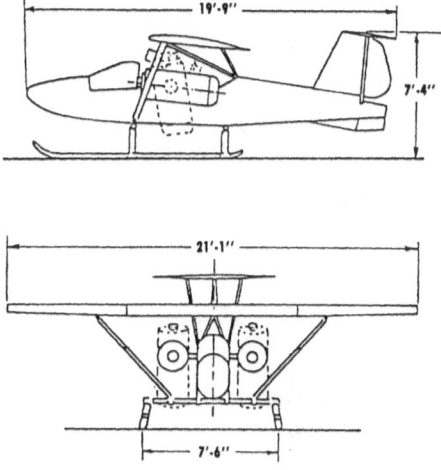

Fig. III-1. Air Test Vehicle (ATV)

Table III-1. Method of Flight Control of ATV

Control Axis	Flight Regime		
	Hover	Transition	Conventional
Pitch	Reaction nozzle at tail	Reaction nozzle and elevator	Elevator
Roll	Reaction nozzles on wing tip	Reaction nozzles and ailerons	Ailerons
Yaw	Reaction nozzles at tail	Reaction nozzles and rudder	Rudder
Hover Height Control		Throttle	
Transition Accomplished		Engine tilt	
Stability Augmentation		None	

Identification: X-14 and X-14A/jet-lift (vectored thrust)

Sponsor/Developer/Time Frame: U.S. Air Force/Bell Aerosystems Co./mid-to-late 1950's

General Information: Open-cockpit low-wing two-seat (side-by-side) deflected-jet aircraft; USAF supported because of ATV success; objectives: horizontal-attitude VTOL and transition capability, repeated flights in VTOL, hover; test instrumentation for full-scale measurements of stability, control and performance parameters

Weight: 3173 lb. (empty)/3300 lb. (gross)

Engine Characteristics: Two Armstrong-Siddeley Viper engines (1750 lb. static thrust each) located side-by-side in nose close to fuselage centerline; upon delivery to NASA, the X-14 was fitted with GE J85 engines and became the X-14A; thrust rotation accomplished by a diversion mechanism substituted for standard tailpipe.

Flight Control System: Hover and transition: jet reaction nozzles at wing tips, nose and tail; RCS powered by compressor bleed air (constant bleed); no artificial stability was provided in basic X-14.

Problems Encountered: First hover attempt failed (engine deficiency and suckdown); HGI problems; operation from elevated platform of perforated steel plate; recirculation of mud and grass led to loss of engine and hard landing; tail buffeting in conventional flight; one accident due to suckdown; another accidnet due to reaction control problem; 10% thrust loss (engine deviation and inlet losses); J85 compressor stalls during VTO (HGI suspected).

Development Status Attained: Flight research.

35

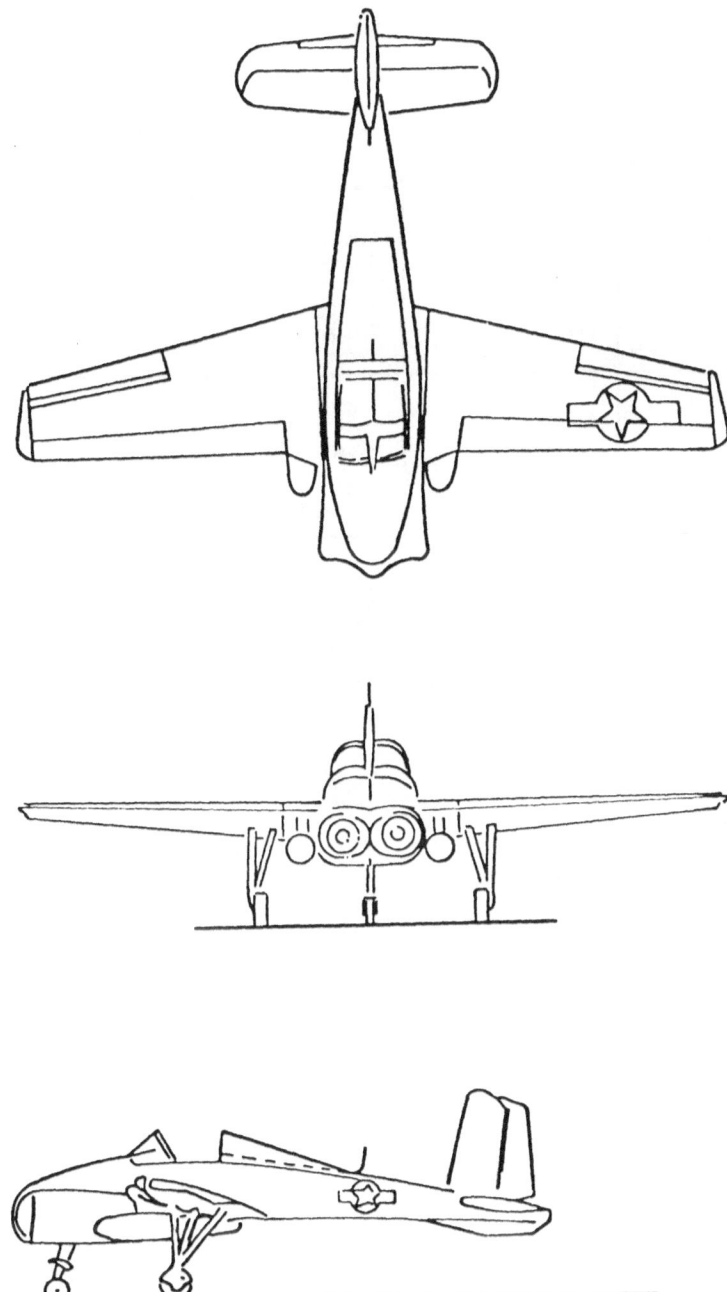

Table III-2. Method of Flight Control of X-14 and X-14A

Control Axis	Flight Regime		
	Hover	Transition	Conventional
Pitch	Reaction jet	Reaction jet and elevator	Elevator
Roll	Reaction jet	Reaction jet and aileron	Aileron
Yaw	Reaction jet	Reaction jet and rudder	Rudder
Hover Height Control	Throttle		
Transition Accomplished	Thrust deflection		
Stability Augmentation	X-14: none X-14A: response-feedback variable-stability system		

Identification: SC.1/jet-lift

Sponsor/Developer/Time Frame: United Kingdom/Short Brothers and Harland/late 1950's

General Information: Single-seat, low wing, tailless, delta aircraft employing separate engines for lift and cruise.

Weight: 8000 lb.

Engine Characteristics: Five Rolls-Royce RB.108 engines (2200 lb. static thrust each), four for lift and one for cruise; four lift engines mounted vertically in a two-by-two arrangement in central fuselage; cruise engine located at an angle in rear fuselage.

Flight Control System: All five engines supplied compressor bleed air (11% of total airflow) to a common duct that provided flow to wingtip, nose and tail reaction control nozzles; pitch and yaw: differential thrusting and swiveling of nose and tail nozzles; roll control: differential thrust of wingtip nozzles.

Problems Encountered: Engine intake temperature distortion; HGI (6% thrust loss); operated from elevated platform and grating over a hole (to reduce HGI and suckdown effects); strong dihedral effect in powered lift led to a dangerous tendency to roll divergency; excessive pilot workload during inbound transition; PIO during transition; accident on one aircraft due to SAS fault.

Development Status Attained: Flight research.

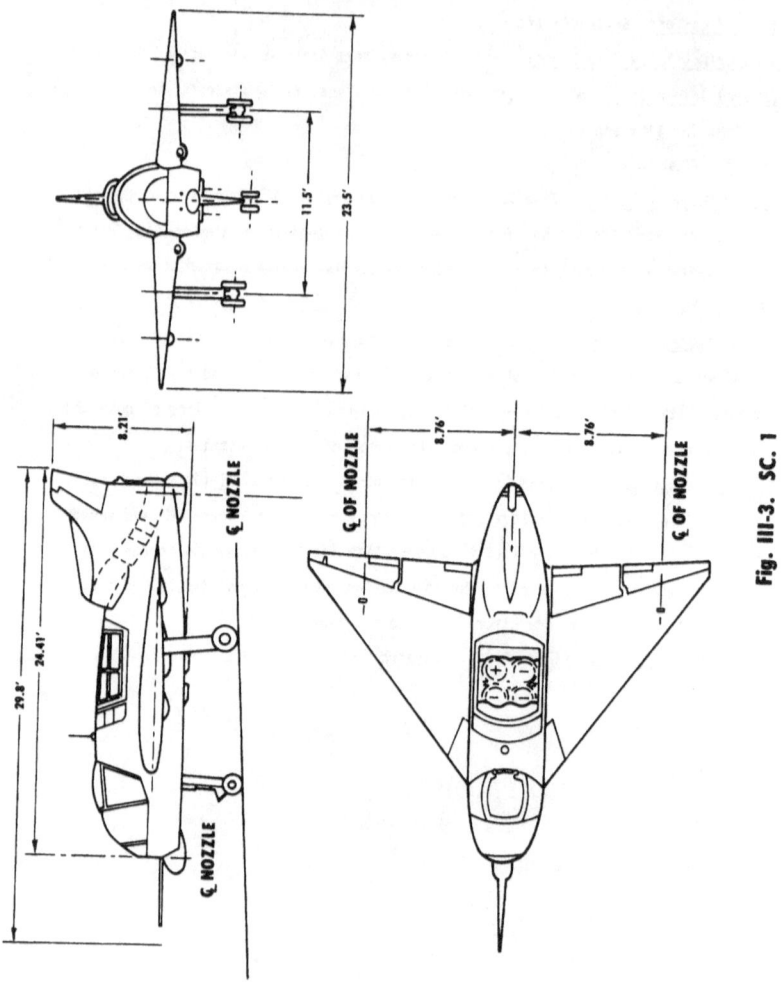

Fig. III-3. SC. 1

Table III-3. Method of Flight Control of SC.1

Control Axis	Flight Regime		
	Hover	Transition	Conventional
Pitch	Reaction jet	Reaction jet and elevator	Elevator
Roll	Reaction jet	Reaction jet and aileron	Aileron
Yaw	Reaction jet	Reaction jet and Rudder	Rudder
Hover Height Control		Throttle	
Transition Accomplished		Lift-engine tilt	
Stability Augmentation		Position controls, rate command, or integrated rate	

Identification: DO-31/jet-lift

Sponsor/Developer/Time Frame: Germany/Dornier/1960's

General Information: High-wing lift/lift-cruise jet transport prototype; unpressurized fuselage contained 30.2 x 7 x 7.2 ft. passenger/cargo compartment with rear loading doors and ramp; NASA flight evaluation (1970).

Weight: 50,000 lb.

Engine Characteristics: Two 15,500 lb. thrust underwing vectored-thrust Pegasus 5 engines and eight 4,410 lb. thrust lift engines in two removeable wing tip pods

Flight Control System: Hover/transition: roll control by differential throttling of lift engines; yaw control by swiveling the lift engines; pitch control by reaction nozzles at tail (reaction nozzles fed by bleed air of main (Pegasus) engines).

Problems Encountered: Major HGI in zero-wind VTO caused main engine surge; reingestion of hot gases in landing; induced forces (lift loss) significant in hover and transition.

Development Status Attained: Flight research

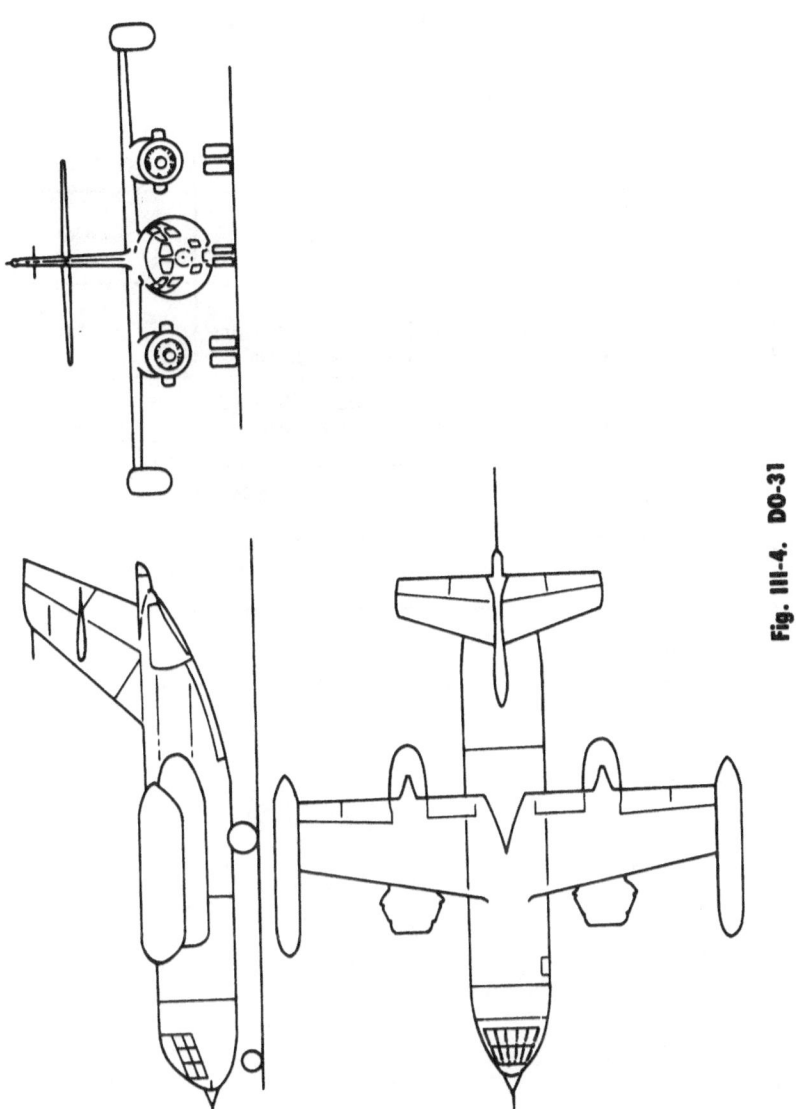

Fig. III-4. DO-31

Table III-4. Method of Flight Control of DO-31

Control Axis	Flight Regime		
	Hover	Transition	Conventional
Pitch	Reaction jets	Reaction jets and elevator	Elevator
Roll	Differential lift-engine thrust	Differential lift-engine thrust and ailerons	Ailerons
Yaw	Swiveling of lift-engine nozzles	Swiveling of lift-engine nozzles plus rudder	Rudder
Hover Height Control		Modulation of main or lift-engine thrust	
Transition Accomplished		Main-engine thrust vectoring	
Stability Augmentation		Yes	

Identification: XV-4B/lift plus lift/cruise (turbojet)

Sponsor/Developer/Time Frame: U.S. Air Force/Lockheed/mid 1960's

General Information: Two-seat (side-by-side) T-tail; used in transition flight investigations in USAF VTOL Integrated Flight Control System (VIFCS) program for developing handling qualities and flight control system design criteria.

Weight:

Engine Characteristics: 6 GE turbojet (3015 lb. static thrust each) - 4 mounted vertically in fuselage and 2 mounted horizontally in nacelles adjacent to fuselage; thrust of nacelle mounted engines diverted downward to combine with thrust of 4 vertical engines to provide lift required for powered-lift flight.

Flight Control System: Powered lift: reaction control bleed air jets for pitch, roll and yaw; transition: reaction jets and aero surfaces; fly-by-wire; mechanical back-up; 3-axis rate damping SAS; dual reaction controls.

Problems Encountered: XV-4B lost in conventional flight accident early in test program; free flight VTOL never accomplished; failures in bleed air ducts due to thermal fatigue; mechanical performance of RCS less than desired; severe empennage buffet induced by downwash; engine stall due to HGI; 1%/3°F degradation in thrust due to recirculation of hot gases; ground erosion, recirculation and tire maintenance problems due to jet downwash temperatures.

Development Status Attained: Flight research

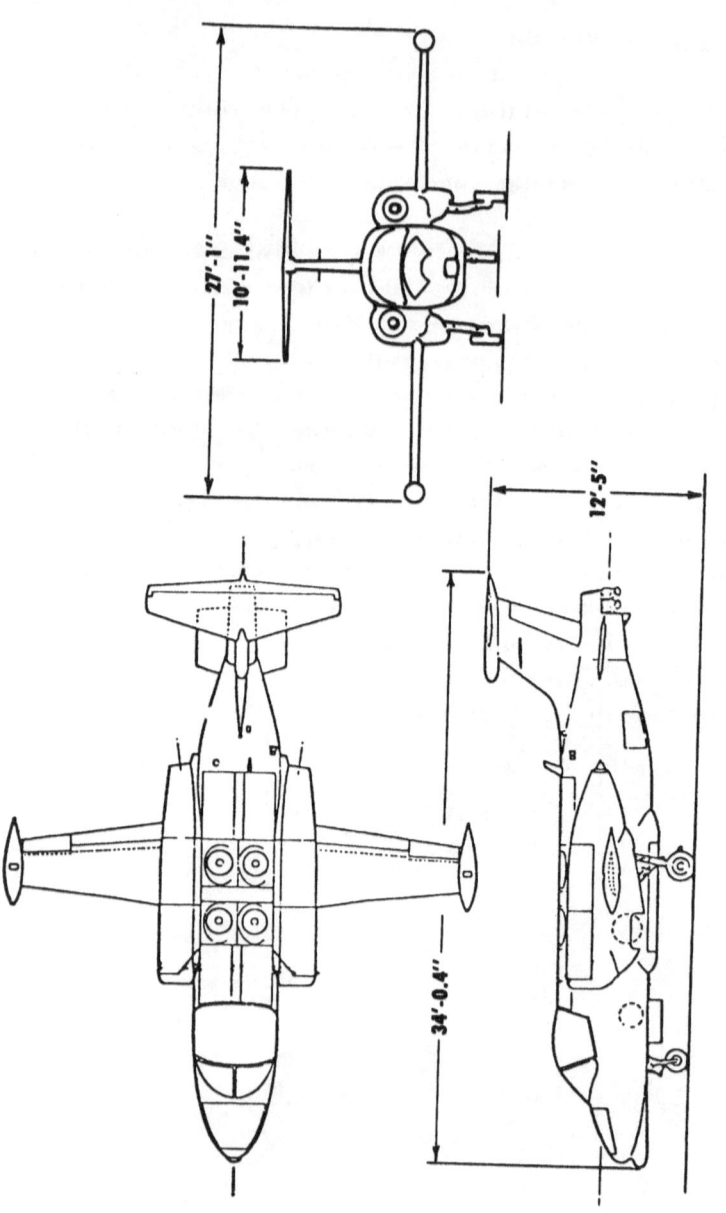

Fig. III-5. XV-4B

Table III-5. Method of Flight Control of XV-4A

Control Axis	Flight Regime		
	Hover	Transition	Conventional
Pitch	Reaction jets	Reaction jets and elevator	Elevator
Roll	Reaction jets	Reaction jets and aileron	Aileron
Yaw	Reaction jets	Reaction jets and rudder	Rudder
Hover Height Control		Engine thrust—throttle or collective	
Transition Accomplished		Rotation of lift-engine nozzles	
Stability Augmentation		3-axis rate damping	

Identification: VJ-101C/lift plus lift/cruise

Sponsor/Developer/Time Frame: Federal Republic of Germany/EWR/mid 1960's

General Information: Supersonic (Mach 2) prototype interceptor; 2 built-the X1 with non-afterburning engines for speeds to Mach 1 and the X2 with afterburning engines for supersonic investigations (the X2 reached supersonic speeds in level flight - a first); the X1 crashed in 1964; originally intended for production; later relegated to research role.

Weight:

Engine Characteristics: 6 Rolls-Royce jet engines arranged in pairs; 2 for lift and cruise in each swiveling wing tip pod, aft of center of gravity; 2 for lift only, in the front fuselage; wingtip pods rotated hydraulically.

Flight Control System: Differential engine thrust (thrust modulation) for pitch and roll; differential tilt of wing pods (in opposite directions) for yaw; three-axis stability augmentation for hover and transition; aero surfaces for conventional flight.

Problems Encountered: Not possible to takeoff absolutely vertical due to exhaust damage to runway; engine failures due to surge; considerable variation in engine response times; structural problems due to high impingement temperatures; far field reingestion; water-cooled grates required.

Development Status Attained: Flight research.

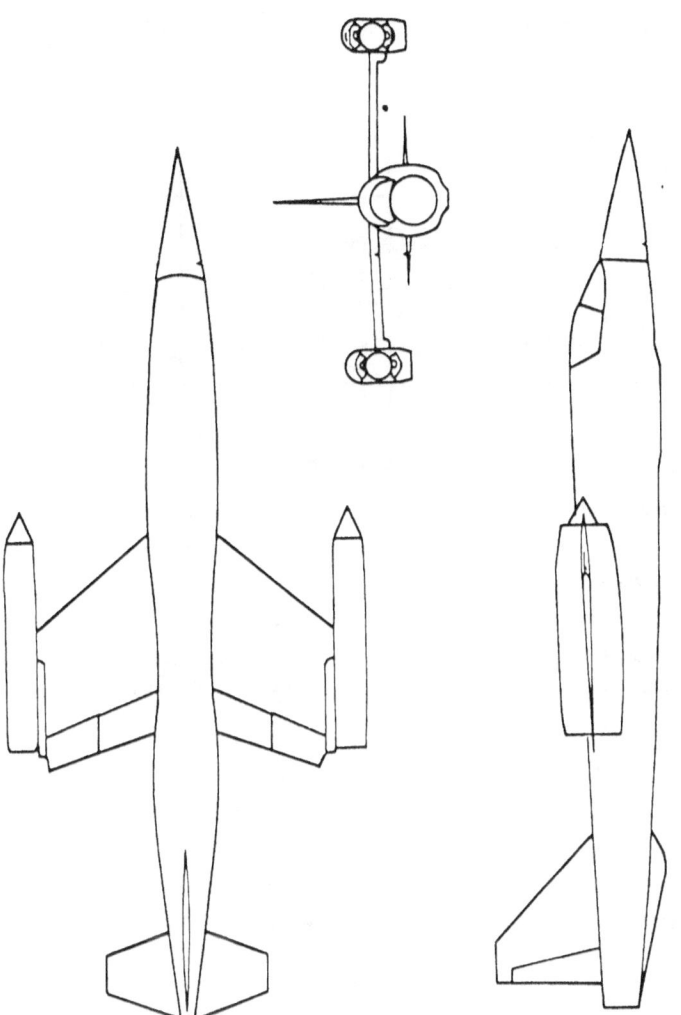

Fig. III-6. VJ-101C

Table III-6. Method of Flight Control of VJ-101C

Control Axis	Flight Regime		
	Hover	Transition	Conventional
Pitch	Differential front/pod engine thrust	Differential engine thrust and elevator	Elevator
Roll	Differential pod engine thrust	Differential engine thrust and aileron	Aileron
Yaw	Differential tilt of wing pods	Differential thrust tilt and rudder	Rudder
Hover Height Control		Throttle	
Transition Accomplished		Engine pod tilt	
Stability Augmentation		Attitude hold in pitch and roll; rate control in yaw	

Identification: VAK-191B/lift plus lift/cruise (turbofans)

Sponsor/Developer/Time Frame: Federal Republic of Germany/VFW - Fokker/early 1970's

General Information: Single-seat lightweight prototype fighter and reconnaissance; NATO requirement; development initially joint with Italy; program changed to reserach aircraft; high-subsonic low-level mission; three prototypes tested; U.S. Navy/FRG joint test program initiated (1974).

Weight: 19,000 lb.

Engine Characteristics: Rolls-Royce, 1 lift/cruise main engine (9094 lb. static thrust) fuselage mounted in center of aircraft, four swiveling nozzles; 2 lift engines (5346 lb. static thrust each) arranged symmetrically fore and aft of main engine.

Flight Control System: Powered lift reaction controls (bleed air) in nose, tail and outer wing panels for pitch, roll and yaw control; triply redundant pseudo fly-by-wire; mechanical backup; conventional flight aero surfaces.

Problems Encountered: Pitch-up in transition; inadequate directional response/damping for precise maneuvering in low-speed powered lift flight; engine thrust 10% less than specified (reduced fuel loads required); main engine susceptible to electromagnetic interference (compressor stall); CTO not feasible; poor VTO overload capability; many "fixes" to solve HGI problems and to improve IGE induced lift; OGE suckdown; nonlinearity in longitudinal control moment; pilot induced oscillations; small low-aspect-ratio swept wing had high stall speed and relatively poor takeoff and landing performance.

Development Status Attained: Flight research.

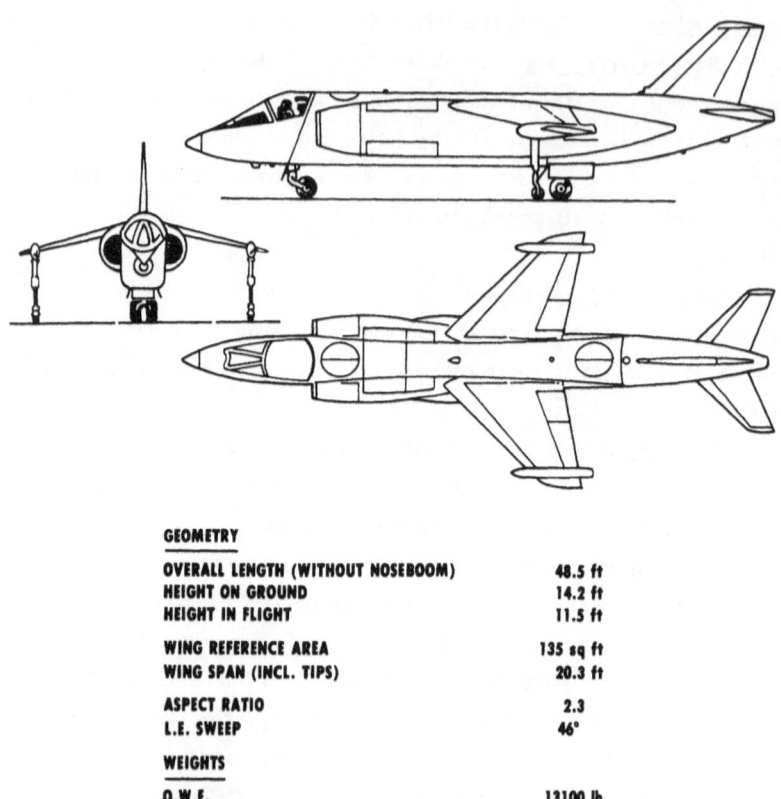

GEOMETRY

OVERALL LENGTH (WITHOUT NOSEBOOM)	48.5 ft
HEIGHT ON GROUND	14.2 ft
HEIGHT IN FLIGHT	11.5 ft
WING REFERENCE AREA	135 sq ft
WING SPAN (INCL. TIPS)	20.3 ft
ASPECT RATIO	2.3
L.E. SWEEP	46°

WEIGHTS

O.W.E.	13100 lb
USABLE FUEL	4200 lb
FLIGHT TEST EQUIPMENT	1410 lb
T.O. WEIGHT	18710 lb

Fig. III-7. VAK-191B

Table III-7. Method of Flight Control of VAK-191B

Control Axis	Flight Regime		
	Hover	Transition	Conventional
Pitch	Nose or tail reaction control	Reaction control and elevator	Elevator
Roll	Reaction control	Reaction control and ailerons	Ailerons
Yaw	Reaction control	Reaction control and rudder	Rudder
Hover Height Control		Throttle	
Transition Accomplished		Thrust vectoring	
Stability Augmentation (Below 120 kt only)		Attitude stabilization in pitch and roll. Rate command and rate damping in yaw.	

Identification: Harrier AV-8A/jet-lift (vectored thrust)

Sponsor/Developer/Time Frame: United Kingdom/Hawker Siddeley Aviation/1958 to present

General Information: Transonic, single-engine, deflected-jet, single-seat, attack/reconnaissance aircraft; shoulder-mounted swept wing; AV-8A evolved from the earlier P.1127/Kestrel (XV-6A); 3000 lb. of ordnance; combat range of 50nm and 360 nm in VTOL and STOVL modes, respectively; operated from 30 ships of various types; in service with U.K., Spanish navy and U.S. Marine Corps.

Weight: 12,200 lb. (empty)/29,000 lb. (gross)

Engine Characteristics: Rolls-Royce axial flow turbofan (21,500 lb. static thrust), counterrotating twin spools to minimize gyroscopic effects; four rotatable exhaust nozzles (two on each side of fuselage); water injection to retain adequate thrust for takeoff or landing under high ambient temp conditions.

Flight Control System: Powered lift: reaction control jets using HP bleed air, jets located at extremities of airplane; 3-axis SAS stability augmentation system, rate damping in pitch, roll and yaw; SAS used only at speeds below 250-kt IAS with landing gear locked down.

Problems Encountered: Deficient thrust of early Pegasus engine; considerable HGI testing on rigs; RC bleed initially marginal IGE; yaw control power too low to hold aircraft in crosswind due to intake momentum drag; combined yaw/pitch RCS redesigned, system made fully variable (demand bleed); tire heating during early tests (melting); random attitude disturbances due to exhaust deflection IGE; $6\frac{1}{2}$% suckdown (strakes solution); fountain interaction effects; high frequency directional oscillations.

Development Status Attained: Operational

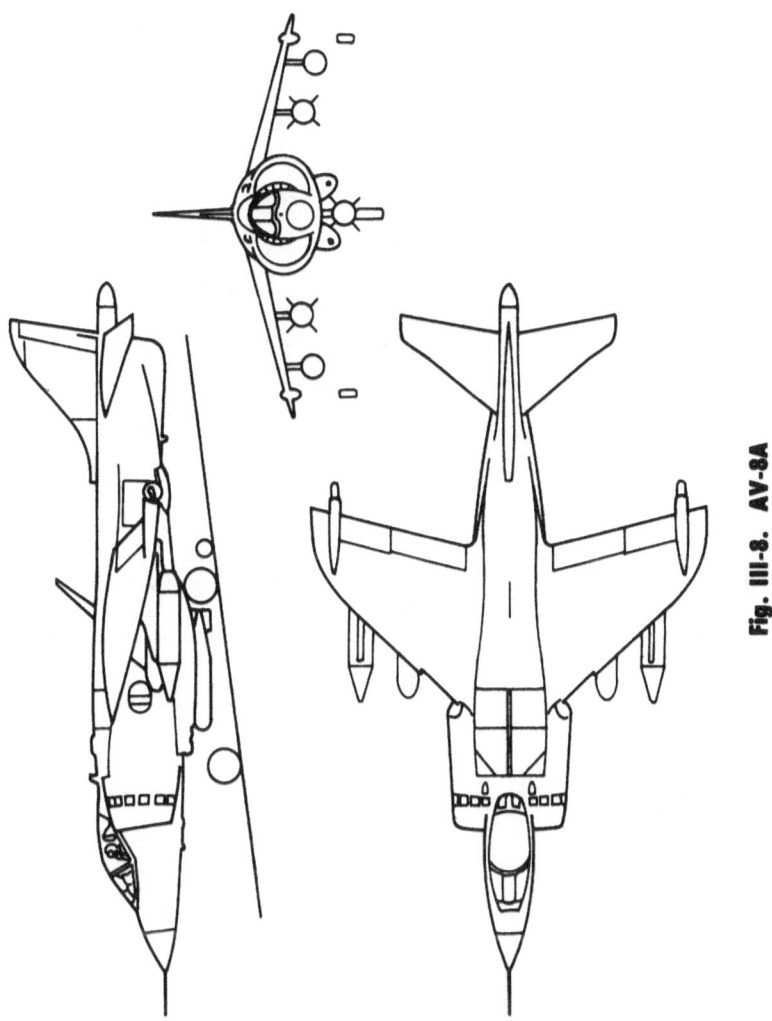

Fig. III-8. AV-8A

Table III-8. Method of Flight Control of P.1127, Kestrel, and Harrier

Control Axis	Flight Regime		
	Hover	Transition	Conventional
Pitch	Reaction jets	Reaction jets and stabilator	Stabilator
Roll	Reaction jets	Reaction jets and ailerons	Ailerons
Yaw	Reaction jets	Reaction jets and rudder	Rudder
Hover Height Control		Throttle	
Transition Accomplished		Thrust vectoring	
Stability Augmentation		P.1127: rate damping, pitch and roll Kestrel: none Harrier: rate damping	

Identification: AV-8B Advanced Harrier

Sponsor/Developer/Time Frame: U.S. Marine Corps/McDonnell Douglas/1973-present

Weight: 12,400 lb. (empty)/29,400 lb. (gross)

Engines: 1 Rolls-Royce Pegasus 11 F402-RR-404 turbofan

General Information:
- improved Hawker Siddeley AV-8A Harrier
- transonic (Mach No. - 0.9+) close-air-support aircraft
- twice the range/payload of today's AV-8A
- augments aerodynamic lift by wing flaps sized and placed to take advantage of engine exhaust flow to considerably increase flap lift
- 7500 lb. of internal fuel
- supercritical wing
    - laminated graphite/epoxy: will weigh 20% less than AV-8A wing
    - 14% greater area than AV-8A wing: more lift
    - positive circulation: more lift
- seven weapons stations (the AV-8A has five) and two gun pods
- inlet
    - redesigned shape and increased inlet area
    - new lip shape and double door auxiliary inlet to increase engine thrust for VTO and reduce distortion during maneuvering
- lift improvement devices
    - retractable fuselage fence
    - fixed longitudinal gun-pod-mounted strakes to capture high energy gases reflected from the ground to increase lift IGE
- stability augmentation attitude hold system: reduces pilot workload
- cockpit improvements
    - cockpit raised 10.5 in. above the AV-8A design to improve pilots field of view rearward
    - larger one-piece wraparound windshield and larger bubbled canopy provides better forward and side vision

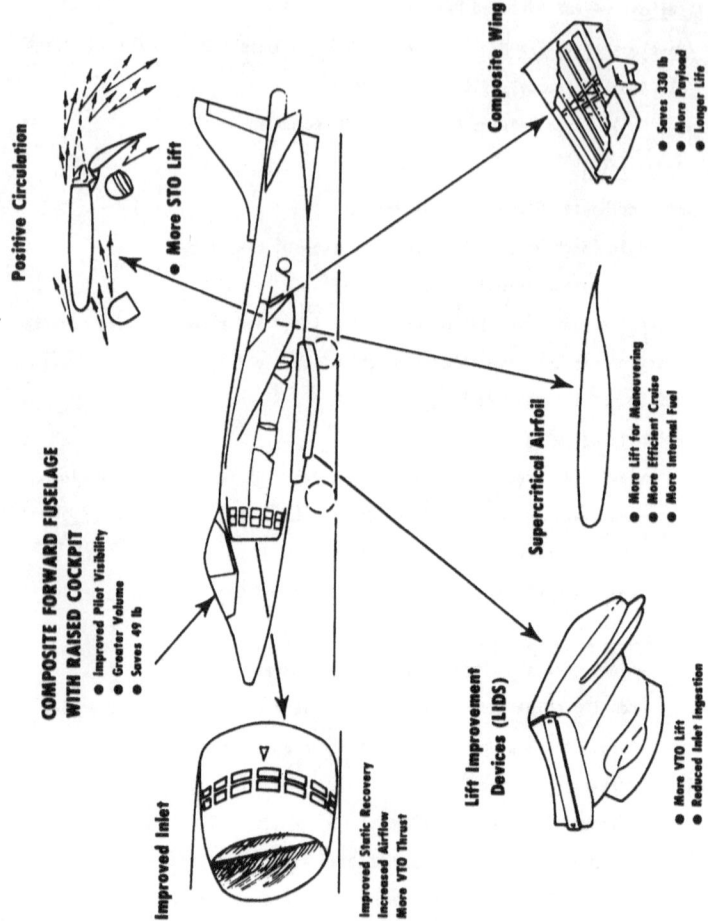

Fig. III-9. AV-8B Technology Improvements

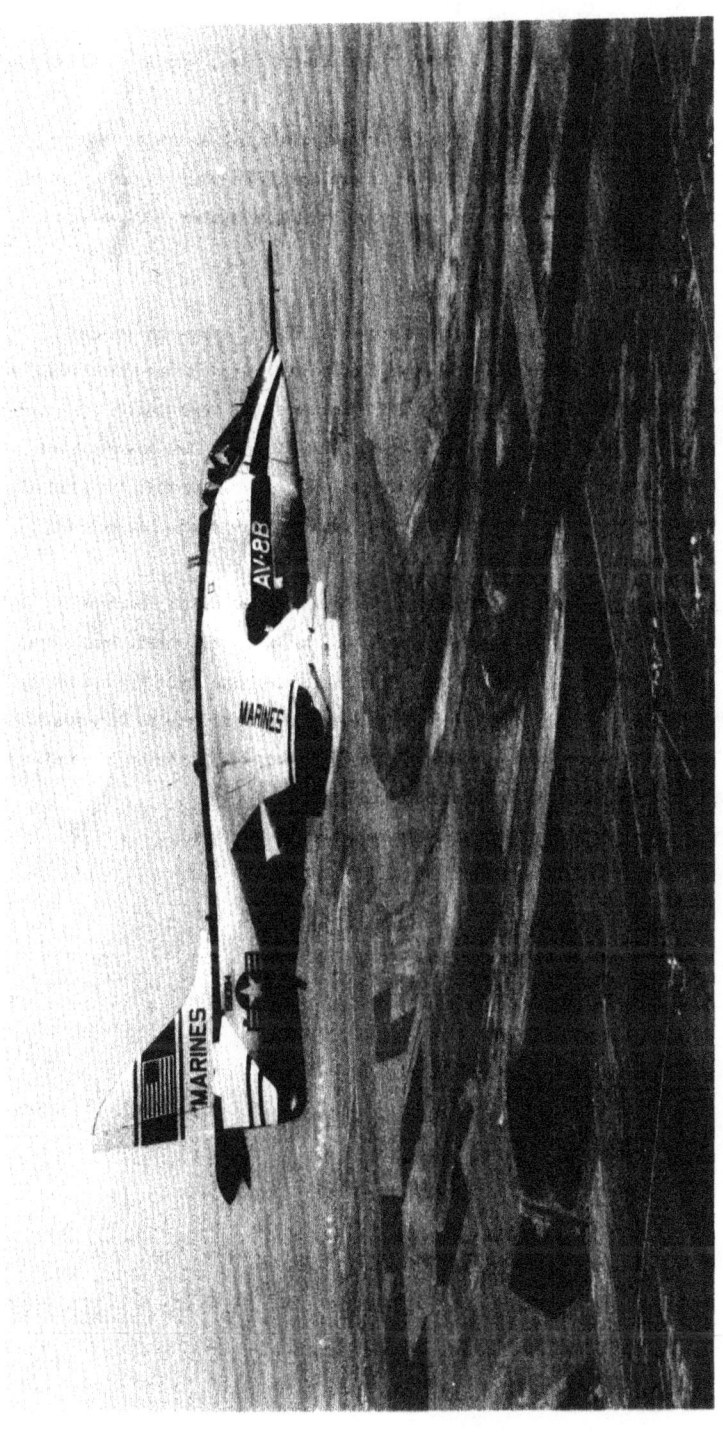

Fig. III-10. AV-8B

Identification: XV-5A/fan-in-wing

Sponsor/Developer/Time Frame: U.S. Army/General Electric Co., Ryan Aeronautical Co./ mid 1960's

General Information: Twin engine, tri-fan, mid-wing research aircraft, turbojet; first fan-in-wing VSTOL; 2 aircraft built to explore flight characteristics of lift-fan propulsion system; GE prime contractor for propulsion system - J85 engines, diverter valves and lift fans.

Weight: 7000 lb. (empty)/9200 lb. (gross)

Engine Characteristics: Two J85 turbojet engines (2500 lb. static thrust each) located in upper fuselage above wing and aft of cockpit; eight-stage rotor, annular combustion system, 2 stage turbine; two lift-fans; one pitch-trim control fan in nose.

Flight Control System: Two primary FCS: one for lift with lift fans operating and one for conventional flight; pitch: spoiling of nose fan thrust; roll: differential stagger of wing fan lowers; yaw: differential vectoring of wing-fan louvers; SAS, 3 axis rate damping.

Problems Encountered: roll control; fan louver hydraulic actuators; unstable in roll at certain landing conditions; ground effects on stability; 15% thrust loss due to HGI (solved by executing VTO tail-to-the-wind, which kept the front fan hot gases away from engine intakes); vertical climb rate extremely sensitive to very small values of excess thrust; strong pitch-up moment with fan start-up; poor hover in ground effect; XV-5A No. 2 crushed, was rebuilt and delivered to NASA as XV-5B.

Development Status Attained: Operational evaluation protototype.

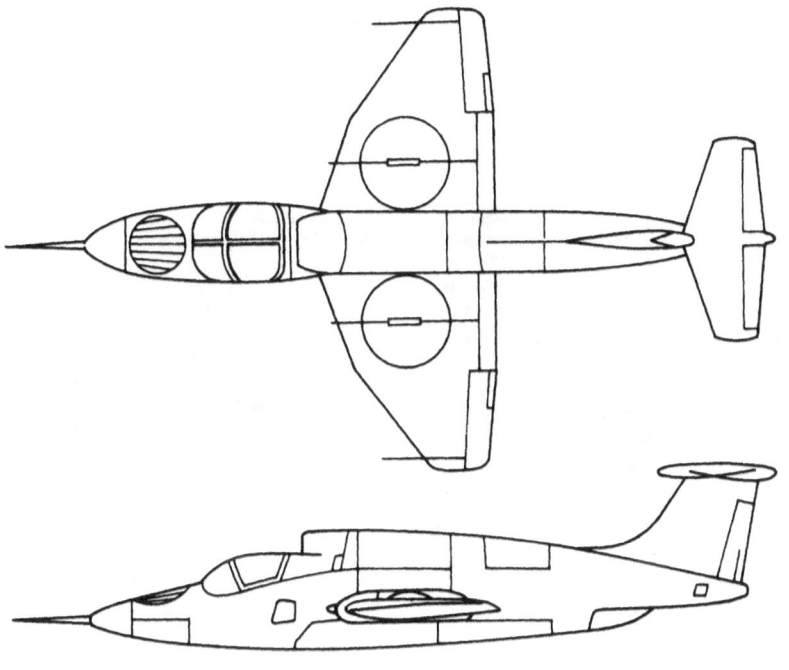

Fig. III-11. XV-5A

Table III-9. Method of Flight Control of XV-5A

Control Axis	Flight Regime		
	Hover	Transition	Conventional
Pitch	Spoiling of nose-fan thrust	Spoiling of nose-fan thrust	Elevator
Roll	Differential stagger of wing-fan louvers	Differential stagger of wing-fan louvers	Ailerons
Yaw	Differential vectoring of wing-fan louvers	Differential vectoring of wing-fan louvers	Rudder
Hover Height Control		Collective stagger of wing-fan louvers	
Transition Accomplished		Vectoring of wing-fan louvers	
Stability Augmentation		3-axis rate damping	

Identification: X-22A/ducted prop

Sponsor/Developer/Time Frame: U.S. Navy/Bell Aerosystems/1960's-present

General Information: Two-place (side-by-side) research aircraft, dual tandem tilting ducted propellers; Navy part of tri-service VSTOL program; ½-size transport vehicle; evaluation of tandem ducted propeller configuration; variable stability system tailored to vehicle; full mil-spec program.

Weight: 14,000 lb.

Engine Characteristics: Four GE 1250 shp T58 turboshaft engines mounted in fairings at root of aft horizontal surface; four 7 ft. diameter ducted propellers driven thru series of interconnecting shafts and gearboxes; power train arranged so that in event of engine failure remaining engines would drive all four props.

Flight Control System: Differential thrust modulation and/or deflection of flow deflectors located in all four ducts; pitch:differential fore/aft propeller blade angle; roll:differential left/right propeller blade angles; yaw:differential left/right elevon deflection; SAS:dual rate damping in pitch, roll and yaw.

Problems Encountered: very high side-force characteristics; transmission and control systems; vibration (blade angle tolerances); fatigue-induced failure on duct leading edge; propeller tip and duct interference; induced moments and fountain impingement IGE; yaw and pitch control IGE; slow response time in hover.

Development Status Attained: Flight research.

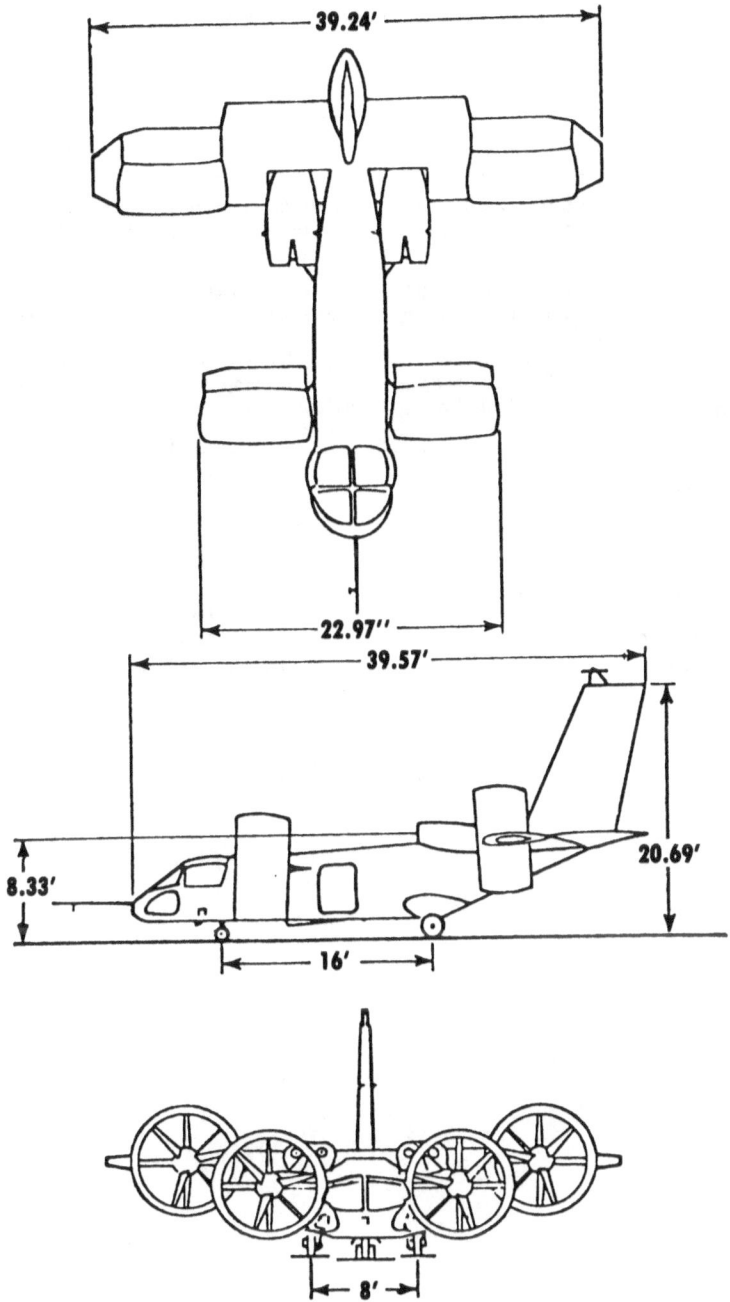

Fig. III-12. X-22A

Table III-10. Method of Flight Control of X-22A

Control Axis	Flight Regime		
	Hover	Transition	Conventional
Pitch	Differential fore/aft propeller blade angle	Differential propeller blade angle and elevons	Differential fore/aft elevon deflection
Roll	Differential left/right propeller blade angles	Differential propeller blade angle and elevons	Differential left/right elevon deflection
Yaw	Differential left/right elevon deflection	Differential propeller blade angle and elevons	Differential left/right propeller blade angle
Hover Height Control	Throttle or collective		
Transition Accomplished	Duct rotation		
Stability Augmentation	Dualized rate damping in pitch, roll, and yaw.		

Identification: XV-3/tilt-rotor

Sponsor/Developer/Time Frame: U.S. Army/Bell Helicopter/1950's

General Information: Tilt-rotor "convertiplane" developed and evaluated as a helicopter; aircraft takeoff and landing, hovering, and low-speed flight were accomplished as a helicopter; conventional surface controls, ailerons, rudder and elevator were used for airplane flight.

Weight: 4800 lb.

Engine Characteristics: Single R-985 Pratt and Whitney reciprocating engine, mounted in the fuselage, provided power to two two-bladed semirigid rotors mounted on swiveling pylons at the tips of a relatively small fixed wing.

Flight Control System: For hover and low speed, longitudinal control was provided by cyclic pitch, directional control by differential cyclic pitch, and lateral control by differential collective pitch.

Problems Encountered: Vibration levels in hover and transition; lateral instability during IGE hover; sudden requirement for large increase in power as hovering flight was approached; excessive blade flapping during conventional flight maneuvering; high drag; weak longitudinal dynamic stability at high conventional flight speed; high-speed rotor and pylon instability.

Development Status Attained: Flight research

Fig. III-13. XV-3

TABLE III-11. METHOD OF FLIGHT CONTROL OF XV-3

CONTROL AXIS	FLIGHT REGIME		
	HOVER	TRANSITION	CONVENTIONAL
PITCH	CYCLIC PITCH	CYCLIC PITCH AND ELEVATOR	ELEVATOR
ROLL	DIFFERENTIAL COLLECTIVE PITCH	DIFFERENTIAL COLLECTIVE PITCH AND AILERONS	AILERONS
YAW	DIFFERENTIAL CYCLIC PITCH	DIFFERENTIAL CYCLIC PITCH AND RUDDER	RUDDER
HOVER HEIGHT CONTROL		COLLECTIVE PITCH	
TRANSITION ACCOMPLISHED		TILTING ROTORS	
STABILITY AUGMENTATION		NONE	

Identification: XV-15/tilt rotor

Sponsor/Developer/Time Frame: NASA-Army/Bell Helicopter/1972-present

Weight: 8500 lb. (empty)/15,000 lb. (gross)

Engines: 2 Avco Lycoming LTC1K-4K turboshaft (1250 shp each)

Speed/Range: 380 mph (max.)/470 nautical miles

Objectives:
- verify rotor/pylon/wing dynamic stability and performance over entire operational envelope
- assess handling qualities and establish safe operating envelope
- investigate gust sensitivity
- investigate effects of disc loading and tip speed on downwash, noise and hover mode operations

Advantages of Tilt Rotor:
- characteristics comparable with helicopter
    - hover
    - low speed agility
    - autorotation
- characteristics comparable with fixed-wing aircraft
    - high lift/drag ratio in cruise
    - low vibration levels
    - high speed agility
    - good reliability and maintainability
    - low noise
    - low vulnerability
- characteristics unique to tilt rotor
    - wide conversion corridor
    - transition smoothly and easily reversible at any point
    - high over-gross-weight capability in STOL mode
    - broad range of flight speeds with engine out

Problem Areas: No significant problems which are fundamental to the tilt rotor concept have been uncovered in the program to date. A basic concern associated with the concept since the flights in the 1960's of the Bell XV-3 tilt rotor aircraft has been the whirl mode instability encountered at that time. All indications are that present design capabilities have successfully resolved that problem and that flight testing will substantiate this prospect. A potential problem area is the possibility that high blade loads could be encountered at high speeds under gust conditions. If loads at high speed are a problem, it should be possible to design the rotor system to accommodate them.

Fig. III-14. XV-15

Identification: XC-142A/tilt-wing

Sponsor/Developer/Time Frame: Tri-service/LTV Aerospace Corp./mid 1960's

General Information: Prototype four-engine tilt-wing; side-by-side seating of pilot and copilot; designed to carry 32 fully equipped combat troops, four litter patients, or 8000 lb. of cargo; five aircraft built.

Weight: 39,000 lb.

Engine Characteristics: Four T64-GE-1 turboshaft engines (2850 shp each) mounted in nacelles under the wing; four-bladed variable-pitch fiberglass propellers; transmission system interconnected the four main props and the tail prop.

Flight Control System: Pitch control:horizontally situated tail prop; roll control: differential main propeller change; yaw control:thru fully immersed two-thirds span ailerons.

Problems Encountered: Stability degradation IGE; STOL performance below expected value; low span efficiency of wing; thrust in hover 12% less than expected; cruise performance was 11% less than expected; conventional flying qualities unsatisfactory; extremely high noise and vibration levels; "greenhouse" effect resulted in cockpit temperatures up to 155°F (refrigerated underwear for pilots was required at Edwards AF Base).

Development Status Attained: Operational evaluation prototype.

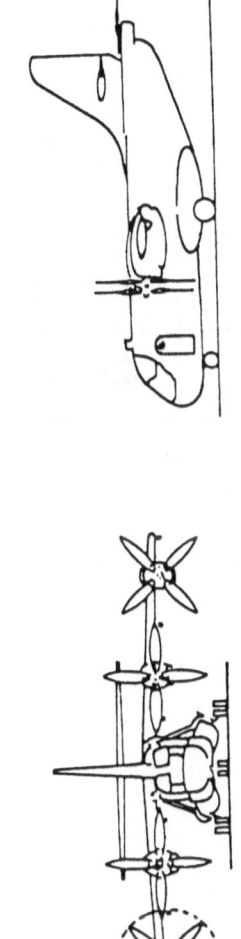

Fig. III-15. XC-142A

Table III-12. Method of Flight Control of XC-142A

Control Axis	Flight Regime		
	Hover	Transition	Conventional
Pitch	Tail propeller pitch	Tail propeller pitch and elevator	Elevator
Roll	Differential tail propeller pitch	Differential tail propeller pitch and ailerons	Ailerons
Yaw	Aileron deflection	Aileron deflection and rudder	Rudder
Hover Height Control		Collective control of propeller pitch	
Transition Accomplished		Wing tilt	
Stability Augmentation		Rate and attitude damping in pitch and roll; rate damping in yaw and height	

Identification: CL-84/tilt-wing

Sponsor/Developer/Time Frame: Canada/Canadair/mid 1960's

General Information: Two-seat (side-by-side) propeller-driven tilt-wing; outgrowth of 7 years of prelim studies aimed at aircraft suited for support of ground forces; studies concluded that tilt-wing was best for VTOL and deflected slipstream best for STOL; CL-84 designed to capitalize on both features.

Weight:

Engine Characteristics: Two 1400 shp Lycoming free-turbine engines mounted in nacelles under the wing; two 14 ft. diameter four-bladed fiberglass main propellers; 7 ft. diameter counterrotating tail propeller located at rear of fuselage.

Flight Control System: Hover, pitch:tail propeller; roll:differential main propeller thrust; yaw:differential flap ailerons (located in propeller slipstream); SAS rate damping about all three axes.

Problems Encountered: Prototype aircraft crashed (1967) on 305th flight after becoming uncontrollable in conventional flight; structural failure of diffuser of one of the engines during transition (cross-shafting supplied remaining torque symmetrically to both props allowing a smooth landing); propeller integrity problems; airframe buffet IGE (due to slipstream deflection); slight lift loss IGE.

Development Status Attained: Operational evaluation prototype; the single CL-84 flew 305 flights and executed 346 VTOL sorties, 109 STOL sorties, 25 CTOL sorties, and 151 transitions; operations from grassfields; hover over water, trees, and personnel; performed rescue tests, including live pickups from land and water.

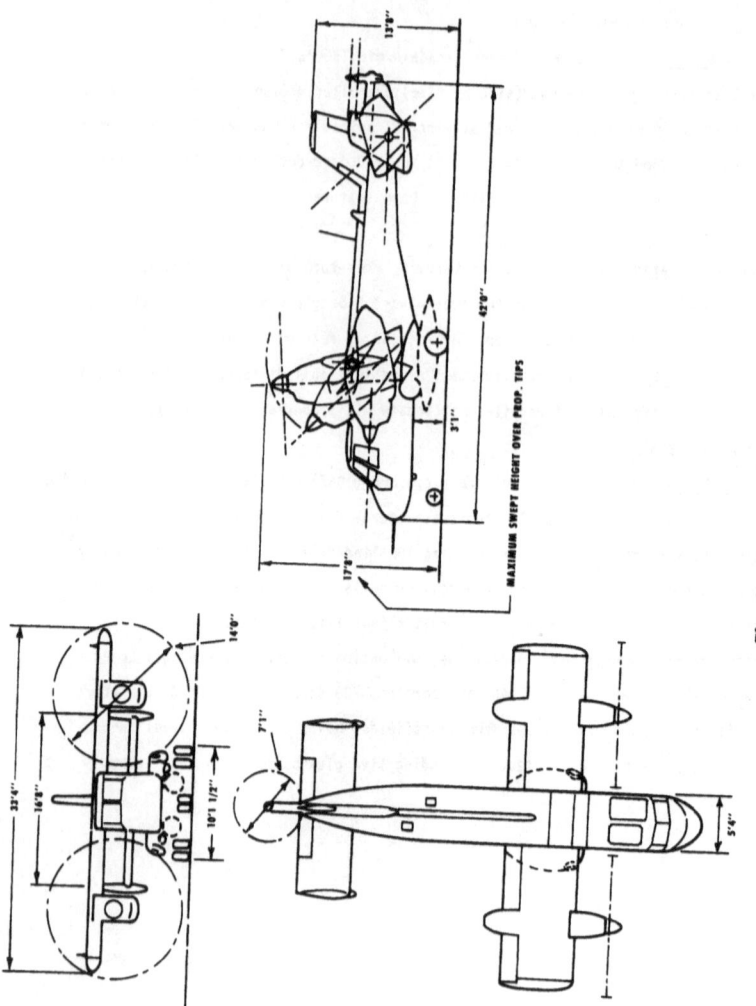

Fig. III-16. CL-84-1

Table III-13. Method of Flight Control of CL-84 Prototype and CL-84-1

Control Axis	Flight Regime		
	Hover	Transition	Conventional
Pitch	Tail Propeller	Tail propeller and elevator	Elevator
Roll	Differential main propeller thrust	Differential propeller thrust and flap-ailerons	Flap-ailerons
Yaw	Differential flap-ailerons	Differential flap-ailerons and rudder	Rudder
Hover Height Control		Power lever	
Transition Accomplished		Wing Tilt	
Stability Augmentation	Prototype	Rate damping in pitch, roll, and yaw plus pitch attitude	
	CL-84-1-	Prototype control augmentation plus yaw	

Identification: VZ-3/deflected slipstream

Sponsor/Developer/Time Frame: U.S. Army/Ryan Aeronautical Co./1950's

General Information: Small high wing monoplane, single-place, single-engine, two-propeller deflected slipstream (wing had downturned tips and very large two-segment area-increasing flaps, that, when fully deflected, boxes in the rear of the wing and turned the propeller slipstream downward to provide vertical lift).

Weight:

Engine Characteristics: 1000 shp Lycoming turboshaft mounted in fuselage; for hover, a swivelling turbine powered nozzle at the tail provided pitch and yaw control.

Flight Control System: Pitch:swivelling turbine exhaust nozzle; roll:swivelling turbine exhaust nozzle; yaw:differential propeller pitch.

Problems Encountered: Could not hover IGE (lateral control was inadequate to cope with ground effects disturbances); no vertical landings were ever made (but STOL performance was outstanding); could not be operated at high angles-of-attack (flow separation on top of wing); two accidents - one due to malfunction in propeller control system and other from a pitch-up and loss of control (pilot error on familiarization flight).

Development Status Attained: Flight research

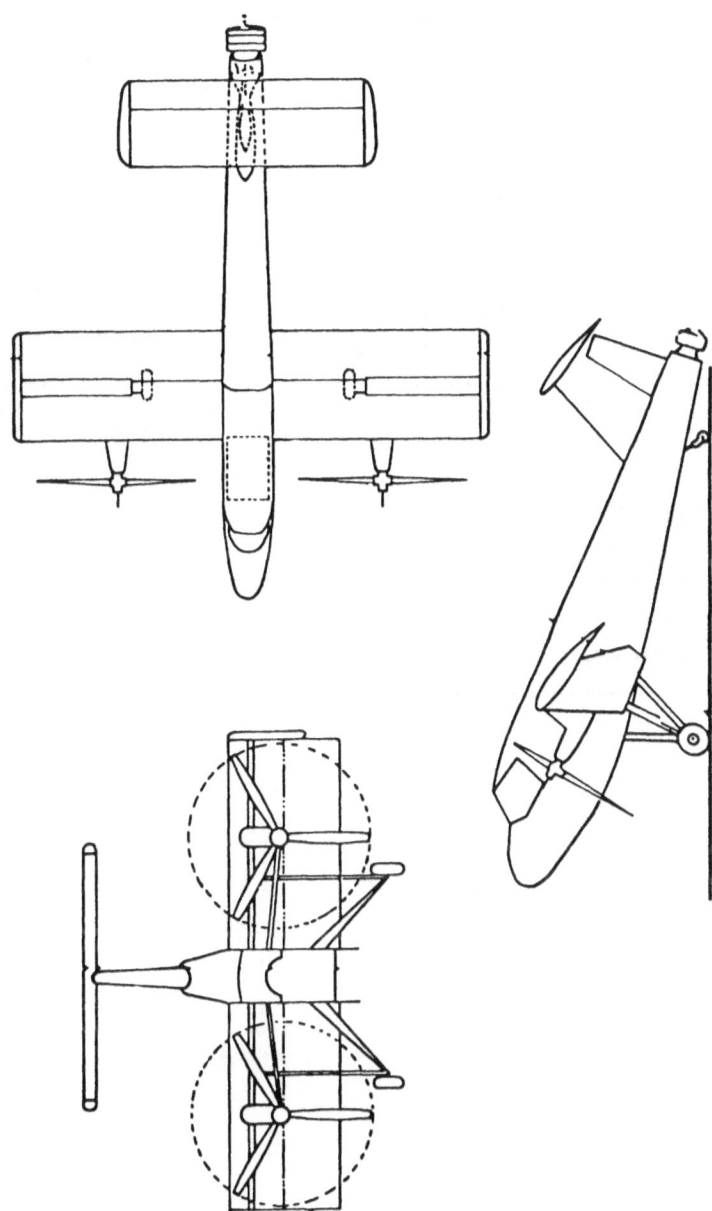

Fig. III-17. VZ-3

Table III-14. Method of Flight Control of VZ-3

Control Axis	Flight Regime		
	Hover	Transition	Conventional
Pitch	Swivelling turbine exhaust nozzle	Turbine exhaust nozzle and elevator	Elevator
Roll	Swivelling turbine exhaust nozzle	Turbine exhaust nozzle and aileron	Ailerons
Yaw	Differential propeller pitch	Differential propeller pitch and rudder	Rudder
Hover Height Control		Throttle	
Transition Accomplished		Attitude Change	
Stability Augmentation		None	

Identification: XV-4A (Hummingbird)/augmenter

Sponsor/Developer/Time Frame: U.S. Army/Lockheed/early 1960's

General Information: Two-seated (side-by-side), twin-engine, T-tail, mid-wing; an outgrowth of early Lockheed jet ejector augmenter work; proposed to Army for Surveillance and Target Acquisition System requirement; funded in 1961 as a proof-of-concept venture.

Weight: 7200 lb.

Engine Characteristics: 2 Pratt and Whitney turbojet engines (3300 lb. static thrust each); exhaust gases directed into jet ejector system in fuselage; thrust augmentation obtained by inducement of secondary flow into ejector system: high-velocity low-mass-flow primary gases mix with low-velocity high-mass flow secondary air to result in momentum increase of mixed gases; jet pump action increased mass flow by about $5\frac{1}{2}$ times.

Flight Control System: Reaction control jets (continuous compressor bleed and exhaust gases) in pitch, roll and yaw; SAS-rate feedback in pitch, roll and yaw; conventional flight:aero surfaces.

Problems Encountered: Designed vertical thrust level not achieved in flight; HGI problems; pitch-up during transition; roll instability in transition; ground effects - aircraft structural vibration caused by flow mixing and back pressure from the ground on augmenter performance.

Development Status Attained: Flight research

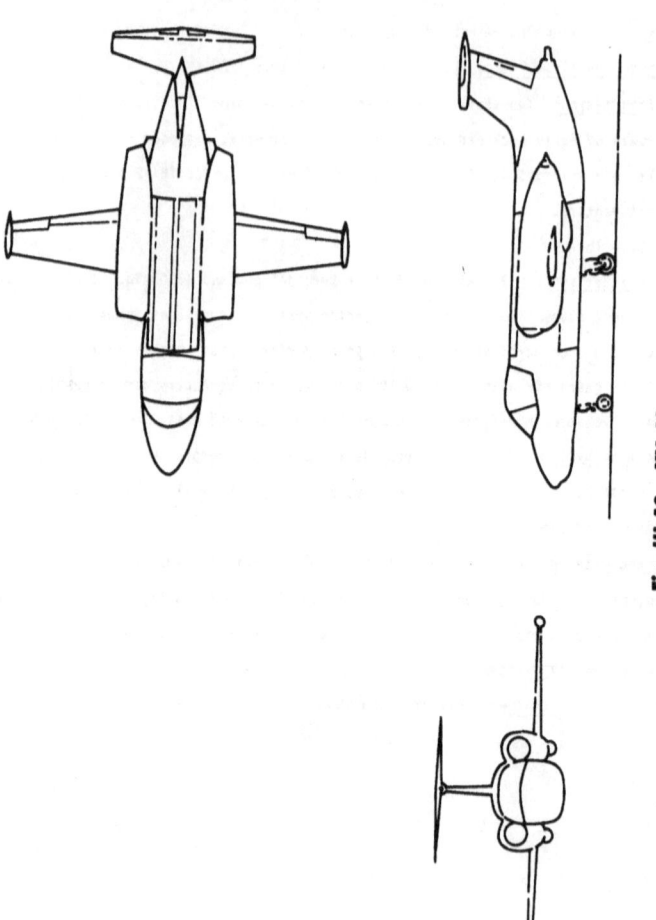

Fig. III-18. XV-4A

Table III-15. Method of Flight Control of XV-4A

Control Axis	Flight Regime		
	Hover	Transition	Conventional
Pitch	Reaction jets	Reaction jets and elevator	Elevator
Roll	Reaction jets	Reaction jets and aileron	Aileron
Yaw	Reaction jets	Reaction jets and rudder	Rudder
Hover Height Control		Throttle	
Transition Accomplished		Aircraft attitude; engine conversion	
Stability Augmentation		Rate feedback in pitch, roll, and yaw.	

Identification: XFV-12A/thrust-augmented-wing (TAW)

Sponsor/Developer/Time Frame: U.S. Navy/Rockwell International (Columbus Div.)/ 1972-present

Weight: 15,000 lb. (empty)/21,000 lb. (gross)

Engines: 1 P&W F401-PW-400 turbofan

Speed/Range: 1,500 mph (max)/ -

Objective: Demonstrate feasibility of a new concept which integrates propulsion, lift, and control in a high performance aircraft

General Information:

An engine two thirds the size required for a direct jet lift concept provides the primary airflow through a diverter system to thrust augmenters located in the wing and canard surfaces. The ejector augmenters are composed of primary nozzles located in a center ejector and above each of the diffuser flaps. Secondary airflow is entrained and mixed with the primary to achieve uniform velocity which effectively increases the installed thrust while maintaining low downwash temperature and pressure footprint. A canard planform provides an optimum augmenter arrangement allowing distribution of vectored thrust. The large amount of overwing airflow induced by the augmentation provides an integrated aerodynamic and propulsion lift which facilitates conversion between hover and cruise flight. The four augmenters are distributed about the aircraft center-of-gravity. Differential modulation of the diffuser flap setting varies the magnitude of the thrust vector from each augmenter providing attitude and height control. Parallel deflection of the diffusers gives thrust vectoring. This eliminates the requirement for a separate reaction system for hover and conversion. The aft diffuser flaps become conventional flight controls when fully retracted for cruise flight. The forward and aft diffuser flaps on the wing are deflected for use as speed brakes. In high-speed flight, use of both the canard and wing flaps for control allows both surfaces to generate a lifting force providing a fighter aircraft maneuvering capability with a minimum of induced drag.

- 35% of the XFV-12A structure comes from existing aircraft: complete forward fuselage and landing gear are from the A-4 Skyhawk, and the inlets, wing box, and main fuel tank are from the F-4 Phantom.
- wings are achieving augmentation ratio of 1.62:1
- canards are achieving augmentation ratio of 1.43:1

<u>Status</u>
- recently began initial tests after repeated delays

Fig. III-19. XFV-12A

Identification: RSRA (Rotor Systems Research Aircraft)

Sponsor/Developer/Time Frame: Army - NASA/Sikorsky Aircraft Co./1971 - present

Weight: 20,812 lb. (empty)/26,250 lb. (gross)

Engines: 2 GE T58-5 turboshaft engines plus 2 TF34-400A turbofan auxiliary engines

Speed/Range: 305 mph (max.)/ -

General Information:
- crew of 3
- 2 aircraft built (now at NASA/Wallops for contractor flight tests)
- crew escape system developed
- designed to fly as a helicopter, as a compound (helicopter plus wing and auxiliary engines), or as a fixed-wing airplane
- flying test bed that will provide a capability to improve rotorcraft prediction methodology and conduct flight research on new rotor concepts
- aircraft are highly instrumented to provide rotor force and moment data
- reduce costly, time consuming trial and error experimentation that has forced development of rotorcraft to proceed at a slow pace
- fly-by-wire control system
- active vibration isolation and measurement system
- electronic flight control system with on-board digital computer to control vehicle during research missions and carry out automatic, pre-programmed maneuvers
- because of its fixed wing and auxiliary engines, RSRA will be able to test rotor systems that might otherwise be too small to support the aircraft or those with unproven control characteristics - the fixed wing also allows the crew to pyrotechnically separate a rotor system in the event of trouble and still return safely to base

Research Tasks: a wide range
- measuring of rotor forces and moments allows detailed evaluation of rotor performance and vibration
- variable-incidence wing, auxiliary thrust engines and drag brakes permit wide range of rotor-operating conditions
- special flight control systems permit rapid, accurate setting of desired rotor-operating conditions
- maneuverability in flight allows determination of dynamic stability derivatives not attainable in any ground test facility

- test speeds - to 300 knots - exceed existing full-scale rotor test facilities
- designed to test variety of rotors
    - teetering rotors
    - articulated rotors
    - hingeless rotors
    - variable geometry rotors
    - rotors with variable diameters or twists
    - slow turning, jet-flap, or circulation control systems

Status

- No. 1 aircraft first flew in October 1976 and has accumulated
    - 23 hr. as a helicopter
    - 1 hr. as a helicopter with the tail used in the compound configuration
    - 2 hr. as a "wingless compound" operating with exterior-mounted TF34 engines and the compound tail, but no wings
- No. 2 first flew in November 1977 and has accumulated 7 hrs. This aircraft has the active vibration isolation/rotor balance system installed
- No. 1 flew in April 1978 as a compound aircraft with auxiliary turbofan engines operating and the wings providing 30% of the lift. It will fly in October 1978 as a pure fixed-wing aircraft with the rotors totally removed.
- The two RSRA vehicles will fly a total of 78 hrs. in the Wallops Flight Center test program before being delivered to the NASA Ames Research Center for rotor development work

Fig. III-20. Rotor Systems Research Aircraft (RSRA)

Identification: ABC (Advancing Blade Concept) Helicopter, XH-59A

Sponsor/Developer/Time Frame: Army-Navy-NASA/Sikorsky Aircraft Co./1971 - present

Weight: 8000 lb. (empty)/10,800 lb. (gross)

Engines: 2 PWC PT6 turboshaft and P&W J60 turbojet

Speed/Range: 160 mph (max)/315 miles

General Information:
- co-axial, counter-rotating, rigid rotor system
- concept shows promise of reducing classical speed limitations of helicopters
- designed to explore flight envelope in pure helicopter mode out to 160 kts and in compound mode with auxiliary propulsion out to 300 kts.
- transition to hover is not required; the throttle settings are changed for forward flight and the faster the aircraft travels the slower the blades rotate, as they operate like wings
- 2 ABC research aircraft have been built

Advantages of ABC:
- improved maneuverability
- improved hover efficiency
- reduced complexity
- deletion of tail rotor
- low noise signature
- compact configuration
- high speed capability

Status:
- Army awarded a contract to Sikorsky for design and fabrication of two ABC research aircraft and flight tests
- 26 July 73 - initial flight of first aircraft
- 24 August 73 - first aircraft crashed due to forward cyclic control requirement exceeding capabilities
- 15 November 74 - new contract awarded which included
    - completing fabrication of second aircraft
    - 50 hr. flight test in pure helicopter mode
    - preparation for, and conduct of, 50 hr. flight test in auxiliary propulsion mode
- 21 July 75 - flight tests resumed with second aircraft with modified control system

- 1976 - the Army determined that funding for high speed (auxiliary propulsion mode) flight tests could not be made available in view of priorities, and solicited interest and support from NASA and the Navy
- June 1977 - Memorandum of Agreement finalized for a joint Army/Navy/NASA program to investigate high speed characteristics through flight tests and conduct full scale wind tunnel tests
- 67 hrs. of flight test in the pure helicopter configuration have been completed
- the aircraft is being reconfigured to add the auxiliary propulsion engines at this time
- 1/5 - scale model wind tunnel tests preparatory to flight tests have been completed

Problem Areas: experience to date indicates three areas in the demonstration aircraft which are related to the basic concept:
- Reduced level of directional control power is available at low pitch settings such as are encountered in autorotation. It appears that this can be resolved in the design for a follow-on vehicle.
- Aircraft structural weight-to-payload ratio is higher than for a single main rotor helicopter. This weight can probably be reduced significantly through the application of composite materials, but the same approach can also be taken with a single main rotor helicopter. The structural weight-to-payload ratio is therefore expected to remain somewhat higher than for a comparable single main rotor helicopter incorporating similar technologies.
- A relatively high hub drag due to large hub. This is yet to be fully explored at the high speeds and may be reducible by means of fairings and other detail design changes. It is planned to explore the affects of adding hub fairings during the full-scale wind tunnel tests at the completion of the flight tests.

Fig. III-21. Advancing Blade Concept Helicopter

Identification: X-Wing

Sponsor/Developer/Time Frame: DARPA-Navy/Lockheed/present

General Information: The X-Wing offers the potential for combining in one vehicle the advantages of the efficient hover performance of the helicopter with the high subsonic speed and long-range capabilities of fixed-wing aircraft. This is achieved by the use of a four-bladed, rigid rotor that can be stopped (with the blades in a 45° orientation to the fuselage) at a forward velocity that allows the aircraft to sustain flight in a fixed-wing mode. The ability to accomplish efficient rotary and fixed-wing flight is achieved by a unique circulation control system blowing air over the edges of the rotor blades which furnishes lift and control for all modes of flight. The circulation control system uses thick hingeless blades with rounded trailing edges. Low-pressure air is pumped into each blade and is ejected from a slot on the upper surface of the trailing edge. This produces a Coanda effect with air adhering to the rounded edge until it reaches the lower surface, keeping the boundary layer from separating and creating high lift. Results of initial design efforts indicate that the dual-plenum blowing, four-bladed circulation control rotor will provide transition capability without dynamic problems.

Advantages

- high payload-to-power ratios
- good hover/low speed handling qualities
- low downwash velocities
- low downwash temperatures
- low noise level
- range three to four times that of standard helicopter

Status

- feasibility demonstrated by six month theoretical analysis and by 1000 hrs. of wind tunnel model testing
- preliminary design of flight demonstration vehicle has been accomplished
- requirements of circulation control system for lift and vehicle control have been defined
- ¼-scale model has been tested to determine the best aerodynamic configuration for transonic flight (Mach 0.8)

- full-scale rotor, hub and flight control system being fabricated for testing in NASA/Ames 40' x 80' wind tunnel in June 78 (to verify the ability to transition from rotary to fixed-wing flight and to provide information needed for proceeding with fabrication of a flight demonstration vehicle)
- decision to proceed with flight demonstration will be made in late FY 78

Fig. III-22. X-Wing

IV. VSTOL TECHNOLOGY ASSESSMENT

METHODOLOGY

This section provides the assessment of the VSTOL technologies. It focuses on the importance of key technological characteristics and in the capabilities to provide design guidance for future VSTOL aircraft. Within each major technology, specific areas have been selected for discussion based on the direct relevance to hover, low-speed and transition flight regimes of VSTOL aircraft.

The primary measure of the status of a technology is its capability to predict the characteristics of future vehicles adequately. By addressing the state-of-the-art in predictive and measurement capabilities for selected technology areas, it is therefore possible to gain an overall assessment of today's technology and to identify technology gaps.

To assess the estimation capability, the following three categories were selected:

- Theory/Semi-empirical
- Empirical: Model
- Empirical: Rig

These categories have been traditionally used to provide design estimates for VSTOL aircraft. The first category includes the applicable theory employed in the initial design as well as the data base of past vehicles and generalized tests conducted within the specific technology area. The last two categories involve the predictive capability of data gathered from tests on similar or identical configurations that are then compared with flight test data. The measurement capability was directed only to flight testing and reflects techniques that have been developed for the in-flight measurement of each of the technology areas.

Predictive capability is rated by the following grading system:

Rating	Definition
none	little or no predictive capability for the actual vehicle characteristics; error (E) between the prediction and the final design is too large to provide even sound engineering estimates: $E \geq 50\%$
poor	predictive capability sufficient for initial estimates but inadequate to formulate detailed design characteristics: $10\% \leq E < 50\%$
good	predictive capability adequate to define final vehicle characteristics: $E < 10\%$

The rationale in support of these evaluations are discussed individually within each technology area in the form of a narrative summary.

The following technology areas were chosen for assessment:

Propulsion and Propulsion Induced Effects

- Inlet Performance
- Bleed Effects
- Engine Thrust Modulation
- Propulsion Sizing
- Environmental Effects
- Lift Generator Performance
- System Losses
- Ground Environment
- Hot Gas Ingestion
- Induced Forces and Moments

Stability and Control

- Equilibrium Trim
- Static Stability
- Control Power
- Gust Sensitivity
- Coupling
- Dynamic Stability
- Height Control

The VSTOL technology matrix is complicated by the burdensome variety of VSTOL aircraft which have been designed, developed and tested over the past thirty years or so. In order to keep the technology assessment within reasonable bounds, the grade for each technology area is averaged over the following seven generic types of lift-generation systems:

- Jet-Lift
- Fan-in-Wing/Body
- Ducted Prop
- Tilt Prop/Rotor
- Tilt Wing
- Deflected Slipstream
- Augmenter

PROPULSION AND PROPULSION INDUCED EFFECTS

INTRODUCTION

A primary distinction between VSTOL and conventional flight modes is the characteristics and conditions of the flow environment induced by the propulsion system. Although the type and severity of these propulsion-induced effects vary widely for different VSTOL concepts, their presence nonetheless requires that major design consideration be afforded to them. All VSTOL vehicles inherently suffer from one problem: the interaction of the propulsive flow with the aircraft. The breadth and depth of this problem are strongly dependent on the disk loading of the aircraft as dictated by its generic lifting mode. Jet-lift aircraft are characterized by severe hot gas ingestion (HGI) and suckdown problems. The deflected-slipstream, tilt-wing, and tilt-duct aircraft are subject to significant forces and moments induced by the slipstream during takeoff and hover, and the fan-in-wing and augmenter aircraft suffer from considerable flow-induced problems associated with the abrupt momentum change of the turning flow in the fans and augmenter. The problems are discussed in this section by addressing the major technology areas considered to be important.

There is a dearth of comparative data on the adequacy of design prediction methods with the final flight vehicle performance. This lack is particularly true in the propulsion disciplines. It is best explained (or rationalized) by the predominance of VSTOL vehicles designed about an existing powerplant. All U.S. and Canadian VSTOL vehicles have used available engines with only slight modifications, if any; thus, the national experience in designing VSTOL engines and assessing the end product is nonexistent. In Europe, the trend is completely reversed, with most of the vehicles using engines designed or extensively modified specifically for VSTOL application. Surprisingly, all of these vehicles were jet-lift aircraft, which require perhaps the most sophisticated technology to match the wide operating envelope. Both the Bristol and the Derby Engine Divisions of Rolls Royce have designed and produced all the engines associated with these vehicles except the Mirage III-V. The lack of available propulsion data to make predictions is very evident particularly for bleed effects, response time, and temperature sensitivity of the engine. This does not mean that engine and airframe manufacturers do not have predictive techniques to provide initial judgments with which to compare to flight engines; it does infer, however, that these data are not readily available.

1. Technology Area - Inlet Performance

    - pressure recovery
    - pressure distortion
    - temperature/pressure distribution
    - inlet momentum drag
    - flow field definition

    Estimation Capability

    - theoretical/semi-empirical: <u>poor</u>
    - empirical
        model: <u>poor</u>
        rig : <u>good</u>

    Measurement Capability

    - flight test: <u>poor - good</u>

    Remarks

    - theory/semi-empirical
        -submerged inlets
        -high flow turning & separation
        -influx/efflux centroid of flow

    - empirical-model
        -Reynolds number effects
        -mass flow scaling
        -influx/efflux simulation

    - empirical-rig
        -full scale duplication of inlet design
        -possibly no influx/efflux simulation

    - flight test
        -inlet conditions to fans
        -propellers and augmenters difficult to measure with flight-worthy instrumentation

Technology Summary

During the development of a new aircraft system, it is incumbent upon the designer to have intimate knowledge of the flow field that will be presented to the engine face. Considerable effort has been expended on several vehicles in both full-scale wind tunnel and rig tests to satisfy the inlet compatibility with the aircraft configuration. There exist no proven methods other than full-scale test rigs that provide an adequate basis for inlet location optimization for an arbitrary vehicle configuration. Sound design techniques and aerodynamic and weight influences of inlet selection should be addressed to provide a realistic preliminary design and to reduce the subsequent necessity of full-scale representation.

The relationship between configurations, inlet location, and flow field effects must be established. A concentrated effort on the development of theoretical and semi-empirical techniques in parallel with wind tunnel tests is necessary to define a good preliminary design and parametric base for future inlet/con-

figuration designs. Wind tunnel tests should include both part-and full-scale
models to provide the necessary flow and model scale laws for timely and suitable
performance prediction. Theory and model results must be verified by using
mobile rigs. The final product should provide proven methodoligies by which the
effects of configuration, inlet design and performance, and flow can be evaluated
for a wide range of potential configuration.

With regard to the operational influence of the inlet on the propulsion and
lift generation system performance, there are no adequate theoretical or
semi-empirical techniques to estimate design levels of inlet losses and distortion
levels. Model tests are also inadequate because of the inability to simulate mass
flow characteristics for influx and efflux flows. The large number of required
inlet performance characteristics for a specific aircraft at the various aircraft
attitudes, speeds, and inflow angles presupposes that a purely theoretical solution
is impractical. Thus, alternate solutions must be addressed.

Methods for estimating compressor/fan/propeller response to both temperature
and pressure distortion must be determined by model and rig testing. Detailed
programs, preferably by an engine manufacturer, are necessary to identify the
governing parameters influencing stall propagation, to define tolerance limits
through predictive techniques gathered from rig tests, and to develop a design
criteria from which future VSTOL engines can be developed. The effects of
bleed, hot gas ingestion, and temperature sensitivity should be combined in
an overall attempt to provide an engine that would be tolerant to the severe
flow characteristics common to VSTOL aircraft.

General

In the development of any new VSTOL vehicle, the designer must consider the
influence of the flow inlets on both the performance of the engine and of the
vehicle. The type of generic vehicle (or actually the type of lift generation
system) ususally prescribes the importance of the inlet flows on the propulsion
engines. In general, the higher the disk loading of the vehicle, the more
critical is the knowledge of the flow field and condition of the air being delivered
to the engine face. Jet-lift, fan-in-wing, and augmenter aircraft and their
inherent high subsonic or supersonic mission goals require that both the main
engine inlets and the auxiliary (secondary) air inlets must not impair the high-
speed performance of the vehicle by their VSTOL features. In addition, there
must be sufficient pressure differential across lift engine inlets for windmill

start of the engines at the beginning of transition; there must be a good flow distribution for engine acceleration demands; and, finally, the intakes must provide the highest efficiency at VTO conditions to assure that the maximum lift-thrust is delivered by the engines.

For the DO-31, a characteristic decrease in inlet efficiency (pressure recovery) occurs between hover and transition since the flow turning angle rapidly approaches 90 degrees as transition is accomplished. The lift/cruise inlet recovery of the VAK-191B offers one means of achieving good inlet performance at both high and low speeds; specifically the use of an auxiliary inlet (sliding shroud) permits flow efficiency from hover through transition and a normal supersonic inlet at higher speeds.

The XV-4B offers the only data available for comparison of the inlet pressure recovery estimation techniques. The comparison shows a large difference between the predicted flow effects and those effects measured on the flight test inlet. The predictive capability over-estimates the pressure losses at all flight speeds and shows a degradation of predicted versus actual pressure losses as throttle demand increases.

In addition to good pressure recovery, pressure distortions at the compressor face must be considered to assure that the engine is not operated in an environment that could induce engine surge. Theoretical and flight test distortion results are available for the hover and transition regions of the VJ-101C. Comparison of test data from several landing and takeoff transitions show that although the theoretical distortion limits were exceeded, no engine stalls were experienced. One concludes that it is difficult to predict exact boundaries by simple distortion parameters.

At this point, it is obvious that the inlet performance, and thus the location of the inlet, is a prime consideration in establishing the vehicle configuration. Considerable testing has been devoted to the inlet performance of jet-lift vehicles, particularly the XV-4B, DO-31, VAK-191B, VJ-101C, and P. 1127. In the other VSTOL configurations, it is not only the main induction system that must be considered but the flow characteristics to submerged fans, augmenters, and ducted and tilted rotors and props located in a diversity of locations on the wings, fuselage, and pods. The duct design for the X-22A required considerable rig testing to prevent duct stall and interference distortion from the tandem duct arrangement. The inflow characteristics to the CL-84 propellers at various tilt angles also required considerable modification to the blade design characteristics to yield adequate performance throughout the flight. For the ducted prop X-22A, the variation of the lengthwise inlet duct area distribution of 1/3-scale model and full scale rig tests show close correlation downstream of the prop, but the contration ratio upstream is considerably overestimated by the scale model.

The effects of inlet design can be far-reaching with respect to other technologies. The designer must be cognizant of the influences of inlet design on the vehicle flying characteristics: the effect of inlet lip shape on wing or fuselage size, the momentum drag and the large subsequent movements induced by changing the momentum direction of the inlet flow, and induced flow impact on the aerodynamic performance of the vehicle. For example, the wing design of the XV-5A was compromised to permit installation of the GE lift fans, and the side flow sensitivity of the XV-4A Hummingbird augmenter inlet limited the flight envelope at low speed. The momentum drag associated with turning the inlet flow can have a severe impact on the trim requirement of the vehicle at all modes of flight but particularly in the transition region. The forces and movements generated by turning the flow are very critical to the stability and control designer because the velocity and power-setting dependent movements must be balanced at all conditions by a judicious mix of reaction and aerodynamic control.

In general, the theoretical and semi-empirical predictive methods are considered to fall short of accurate design capability primarily because of the difficulty in describing the flow field presented to the inlet and the losses and induced forces generated by the turning of the flow. Part-scale model tests provide a somewhat better capability, but the problems of separating aerodynamic and propulsion-induced effects and adequate simulation of the engines and thrust/lift generators are the major drawbacks. Full-scale rig testing is considered the best predictive device for inlet performance and, in most instances, rig test results are used as the basis for flight test calculations.

In-flight measurement of inlet performance parameters is made difficult by the instrumentation installation requirements. The number and location of measurement probes has a significant effect on data quality. The installation including wiring must be structurally and thermally compatible with the operating environment, which usually results in a limited number of probes at key locations, perhaps other than the desired station, to serve as a guide for comparison with rig empirical results. Because of the importance of engine inlet performance on overall airplane performance throughout the wide speed range of jet-lift vehicles, this area usually receives attention in the flight development program of these vehicles. For example, the P.1127/Kestrel/Harrier had an extensive series of intake designs to achieve a satisfactory inlet. Although aerodynamically functional, the inflatable rubber bag lip of the P.1127 was replaced by an all-metal inlet because of problems in getting a smooth suckdown for conventional flight. The Kestrel's metal inlet was substantially redesigned to include blow-in doors to provide for the larger mass flow requirements of the uprated Harrier engine.

2. Technology Area - Bleed Effects
   - propulsion system response to bleed demands
   - interaction of aircraft with reaction control system
   - coupling
   - lift/thrust losses

   Estimation Capability
   - theoretical/semi-empirical: none
   - empirical
     model: poor
     rig : poor

   Measurement Capability
   - flight test: poor

   Remarks
   - theory/semi-empirical
     -transient modeling difficulty
     -compressor/turbine performance maps required
     -point design at best
   - empirical-model
     -transient modeling difficulty
     -inertial scaling effects
     -hot air supply problems with force balance
   - empirical-rig
     -provides full range of dynamic response
     -good for engine response characteristics
     -must estimate aircraft interface
     -poor for effects based in component engine matching or aircraft interface not simulated
     -difficult to isolate bleed effects from total aircraft and engine behavior

   Technology Summary (see Engine Thrust Modulation)

3. Technology Area - Engine Thrust Modulation
   - engine response rate
   - aircraft/engine control blending
   - synchronization

Estimation Capability

- theoretical/semi-empirical: none
- empirical
    model: poor
    rig : poor

Measurement Capability

- flight test: good

Remarks

- theory/semi-empirical: same as bleed effects
- empirical-model: same as bleed effects
- empirical-rig
    -engine performance: same as bleed effects
    -thrust performance directly affected by aircraft response
    -effects reduced as disk loading decreases

- flight test
    -engine and aircraft responses readily determined
    -thrust computation dependent on test stand (rig) data

Technology Summary

The influence of bleed air extraction and engine/thrust response are addressed together since both involve vehicle attitude control. It is recognized that their individual impact on the engine can be extremely different.

The jet-lift and augmenter aircraft with bleed-air control systems experience severe hover and low-speed control problems associated with the inability to maintain sufficient balanced responses to the aircraft during critical conditions. VSTOL aircraft require from 8 to 16 percent of the total airflow be reserved for attitude control purposes, in many cases on a part-constant/part-demand basis. Future aircraft are expected to require as much as 20 percent of the flow during critical maneuvers. Efficient accommodations for bleed control schemes must be sought.

With regard to the aircraft, the tradeoff of constant bleed as opposed to a demand system or a suitable combination of both types must be investigated early in the aircraft design cycle. Potentially high payoff of advanced concepts in bleed air control such as bleed-and-burn, ejectors, and wing-tip fans should be addressed in research programs to provide a more efficient vertical thrust recovery from the control system and to minimize the height coupling that is characteristic of these systems. The possibility of utilizing existing hover and ground test rigs simulating various aircraft inertias as a full-scale testbed for exploration of promising concepts and the determination of control and aircraft response characteristics must be considered as a reliable and primary source of future design criteria. Likewise, reaction control sizing

requirements should address alternate solutions. Flight test experience on the
Y-14 indicated that only a small amount of activity was required for controlling
the aircraft attitude with maximum demands for only short (5 sec.) periods of time.
For the AV-8A, similar tests showed that maximum bleed control was required for
only a relatively short time and could improve considerably with pilot technique.
In an effort to avoid "over-design" of a reaction control system, investigations
should be conducted for a system design that provides less than the maximum
control expected but supplements the additional bleed air for the short-time
peak control demands by operating the engine above limit temperatures and speeds.

Aircraft control through thrust modulation systems employing either direct
throttle control, flow proportioning devices, or differential blade pitch have
imposed considerable development challenges on past systems, most notable of
which are VJ-101C, XV-5A, VAK-191B, X-22A, and CL-84. The utilization of these
techniques requires an intimate knowledge of the dynamic interplay between the
control concept and aircraft response to provide meaningful engine response
criteria. There exist no semi-empirical techniques to provide these criteria,
and model tests have been shown to be inadequate because of lack of adequate
engine simulation. Since completion of the rig and flight testing on these
vehicles, only limited full-scale rig tests investigating these effects have
been uncovered.

For the jet-lift aircraft utilizing a differential throttle control system,
only a full-scale rig test simulating the specific aircraft provides the
needed engine response characteristics. With regard to simultaneous, multiple-
engine flow management, thrust proportioning concepts such as the energy
transfer and control schemes advocated by NASA/McDonnell show considerable
promise as an acceptable attitude control device with minimum altitude coupling.

Flight experience has shown that the thrust modulation systems, which
inherently have their control units widely separated for increased response
capability, require close monitoring of the propulsion and lift generation
systems to assure that lifting forces are within specific tolerances. The
VJ-101C experienced trim and matching problems on the wing-tip engines;
the XV-5A, variations in flow proportioning to the fans; and the X-22 and
CL-84, blade twist variation among the propellers. The existence of these
problems must be considered early in the development cycle of a vehicle to
provide a balanced and predictable force balance.

General

Present VSTOL engines supply as much as 20 percent of the engine massflow to the control system often as a demand ("use it when you need it") or part-demand/part-steady system. The effect of a large demand, particularly from a control system requiring high response, influences the dynamic stability of the engine. It induces rapid changes that could stall the compressor downstream of the bleed ports. The utilization of compressor bleeding incurs an immediate loss of lifting energy supplied by the gas generator; even though a well-designed Reaction Control System (RCS) will minimize this thrust loss by producing positive lift at the control jets, the overall efficiency of the propulsion system as a lift producer is decreased. The attitude/altitude coupling prevalent in all the demand bleed aircraft gives full bearing to the importance of this design feature. Why not a steady bleed system then? For the high rates of bleed required to control the upcoming VSTOL aircraft, the thrust penalty associated with extracting a constant bleed flow is prohibitive in increasing the aircraft size. Thus, the RCS is designed to provide some lifting capability, and the same attitude/altitude coupling is evidenced as in the X-14.

Several vehicle concepts, covering the spectrum from rotors to jet-lift, utilize thrust modulation as the primary controlling element for the vehicle in hover and low-speed flight. Instead of the control jets utilizing bleed air, the pilot or automatic control system directly controls the throttle or lift-generator mechanism as in the thrust vectoring system of the VJ-101C or the energy transfer system of the XV-5A.

To obtain efficient overall performance, the engines must meet certain demands concerning the response time and synchronization of the control system. As in the bleed discussion, thrust modulation control directly affects the control power sensitivity and attitude/altitude control by the efficiency of the engine control schemes. Since vehicles utilizing thrust modulation for control are inherently multi-engine aircraft, there can be significant changes in the forces and moments induced by the changes in the interaction of the flows on the vehicle in hover conditions. Modulation must be achieved without adverse effects on vehicle attitude by altitude-hold engine/flight control schemes. Extreme care must be exercised continually to assure uniform response by the system. As an example, slight variation in collective pitch response from each engine of a tilt-rotor aircraft would produce an instantaneous moment that could impair takeoff or, even more dramatically, place the aircraft in an attitude from which recovery cannot be accomplished. In short, both bleed and modulation systems used for control of VSTOL vehicles during vertical, hover, and low-speed flight must be integrated and blended with the aerodynamic controls so that pilot response is natural during all flight modes and performance penalties are minimized.

In-flight measurement of bleed airflow is impractical, and direct measurement of thrust is impossible for any of the generic types. These facts, combined with the complexity of the design for the bleed and thrust modulation system for the specific application, require detailed rig testing on the vehicles invovled and, in several cases, on a hover rig. Thus, the complele lack of comparative data between design and flight test is not surprising for these two disciplines.

4. Technology Area - Propulsion Sizing

- thrust augmentation concepts
- overspeed
- water injection
- VTO and emergency ratings

Estimation Capability

- theoretical/semi-empirical: good - poor
- empirical
    model: not applicable
    rig : good

Measurement Capability

- flight test: not applicable

Remarks

- theory/semi-empirical
    -good for performance prediction
    -thermo-physical relationships apply
    -small data base

- empirical-rig
    -extended service testing possible
    -cyclic testing

Technology Summary (see also Lift Generator Performance)

Propulsion system sizing considerations form a significant design constraint on VSTOL vehicles. The inherent high power ratio required for verticle lift and the influences of high ambient and induced inlet temperatures necessitate an oversizing of the propulsion system as well as its accompanying weight and volume penalties. The desirability of establishing a VTO or short-lift-type rating is a foremost consideration to realize potential gains in performance from future aircraft designs. Several investigations have been conducted to evaluate the potential merits of VTO rating by various schemes, e.g., water injection, plenum chamber burning, overspeed. However, the data base of operating experience is low for these special ratings with only the Pegasus engine having been designed initially for operation in this manner. Tests on various engines have shown that the engine characteristics during increased rating can be predicted adequately from thermodynamic considerations and desired vehicle performance can be achieved by extending the takeoff rating of conventional engines used for VSTOL, e.g., the XV-4B. The primary concerns with the establishment of VTO ratings are the effects of ambient conditions on the engine during these periods of operation and the lack of firm governmental position on the allowance of these ratings during initial design consideration.

Parametric tests should be conducted to determine the effects of high ambient temperatures and temperature transients on the performance characteristics of the engine for both normal takeoff and special VTO ratings. These tests should provide an adequate baseline from which to assess the controlling criteria for stall-free flight, i.e., effect of temperature transient on surge margin, overtemperature limits. A policy should be established permitting the use of a short-term (at least 2 minutes to cover takeoff and transition) rating in future VSTOL aircraft to realize the design potential of the vehicle and to induce the U.S. engine manufacturers to consider in detail the effects of increased ratings on engine life.

5. Technology Area - Environmental Effects
    - engine temperature sensitivities
    - dynamic propulsion interactions

   Estimation Capability
    - theoretical/semi-empirical: good - none
    - empirical
        model: not applicable
        rig : good

   Measurement Capability
    - flight test: good - poor

   Remarks
    - theory/semi-empirical
        -good for uniform temperature variation
        -performance based on thermodynamic considerations
        -none for temperature distortion or temperature transients
        -usually more critical as disk loading increases
        -function of compressor or stall margin

- empirical-rig
  - controlled environmental testing possible
- flight test
  - good for uniform temperature field
  - temperature distortion effects difficult to measure, but generally not a problem with low disk loading vehicles

Technology Summary (see Lift Generator Performance)

6. Technology Area - Lift Generator Performance
   - augmenter/ejectors
   - thrust/lift
   - propeller/duct
   - aerodynamic turning

Estimation Capability

- theoretical/semi-empirical: poor - none
- empirical
  - model: poor - good
  - rig : good - poor

Measurement Capability

- flight test: good

Remarks

- theory/semi-empirical
  - installation effects primary factor
  - aircraft/propulsion flow interactions uncertain particularly for lower disk loadings
  - transition performance particularly difficult
- empirical-model
  - simulation of jet and fan flow characteristics
  - ground proximity suppression effects scaling laws uncertain
  - inconsistent parameter scaling
- empirical-rig
  - good static performance prediction
  - limited value for transition performance
- flight test
  - thrust computation procedures vary with generic types and are dependent on empirical inputs

Technology Summary

Propellers/Rotors - For concepts employing propellers/rotors for both lift in hover and propulsion for cruise, high levels of performance are necessary to produce an efficient vehicle in both modes of flight. The determination and optimization of propeller cruise performance have received considerable attention; unexpectedly, static thrust performance has not been considered, so that a serious gap in reliable static performance estimating techniques has resulted. The complicated lift load distribution over the propeller blades induced by strong tip vortices, the complex inflow to the prop/rotor tips, and the influence of a wing/body on the contraction of the slipstream are major problem areas. For the ducted propeller concepts, static conditions can be estimated with an acceptable degree of confidence but the effects of forward speed on the duct/propeller combination have not been resolved.

Installed prop/rotor performance is influenced directly by the recirculative flows and/or body interaction; to date, empirical comparative estimation techniques offer only a poor capability at best.

Static performance estimation and the influence of ground proximity on prop/rotor performance must be addressed. There is a need for parametric studies of varying arrangements of lift generator installation relative to the other major vehicle components (i.e., wing, fuselage) to investigate performance sensitivities to vehicle geometries. The scaling laws applicable to model and full-size configurations must be determined.

In addition, investigation of propellers as the main source of aircraft pitch control by use of monocyclic blade pitch control is worthwhile since it provides the attraction of eliminating the tail rotor and associated cross shafting/transmission/control problems. Extensive control dynamics studies including theory and model tests are needed to evaluate all potential modes of control and to provide ultimate design criteria. Additional influences are imposed on the propeller blades resulting from the cyclic blade changes and the complications that will invariably arise from slipstream and wing/body interferences.

Augmentation - Ejector technology requires considerable attention. Although a sound principle, it is apparent from the difficulties in attaining augmentation ratio goals, as witnessed by the XV-4A program and the poor correlation in predicting augmentation from laboratory models, that some technology improvement is in order. In spite of continued research in this area, insufficient knowledge of the interactions between the primary and secondary flows remains. Although the Navy has been intimately involved in augmenter wing aircraft development, the effort to understand the basic internal aero/thermodynamics of the mixing process has not been sufficiently supported by in-house programs. The relative success of ejector/augmenter performance to date is a combination of art and science. The fundamental knowledge of viscous mixing, wall phenomenon/end wall geometries, and pressure/temperature effects have been shown to be the major uncertainties in ejector performance. The marriage of an ejector system with an aircraft application (i.e., XFV-12 and XV-4A) has indicated that the external flow field manifests a significant influence on the overall lifting/propulsive reaction of the augmenter.

The Navy should reinstate the research in augmenter/ejector technology initiated by ONR almost 20 years ago to provide the tools and techniques necessary for adequate application of this promising technology to aircraft. Consideration should be directed toward development of the theoretical/semi-empirical techniques to describe losses and viscous flow interactions affecting ejector performance. Utilization of model tests (preferably in three-dimensional configurations) and the rotary rig at Rockwell International would enable parametric studies to be conducted to describe the effects of pressure ratio, temperature, augmenter shroud geometry, and in-flow patterns on overall performance. Particular attention should be concentrated on the development of scaling laws for model/full-scale performance estimation and the effects of ground proximity on performance.

Lift Fans - The attractiveness of fan-in-wing and fan-in-fuselage designs for their low downwash velocity, potentially low-temperature profile and decreased sensitivity to hot gas ingestion indicates that a more thorough understanding of the advantages and limitations of this propulsion device is required. For aircraft application, the dependence of fan pressure ratio, fan location, and aircraft configuration on overall performance cannot be predicted adequately. Present scaling laws have not indicated that a satisfactory level of confidence exists for proper model/full-scale simulation.

The relative ease in simulating fan flow can be exploited by conducting powered model tests to determine accurate empirical methods of predicting flow field parameters (e.g., induced flow over the wing/fuselage, forward speed trim effects) and applicable scaling laws. With respect to fan technology, investigation of the advantages of twin-scroll, higher pressure ratio fans should be conducted to provide the balance between engine/fan complexity and increased aircraft performance. Continued applied research to flow management of fans (e.g., the energy transfer and control (ETAC) schemes advocated by NASA/McDonnell) and a parametric study to identify the potential of utilizing existing engines with projected fans to provide alternative aircraft performance tradeoff with jet-lift should be considered. Variable pitch fans offer an attractive potential for high-response aircraft control; the development of analysis methods to determine performance/weight/cost tradeoffs as well as the suitability of the driving mechanism (i.e., direct gas flow or mechanical drive) need improvement in design and performance estimation.

Jet Lift - The influence of jet engines on the surrounding flow field becomes increasingly dominant as the freestream velocity approaches hover. At present, there are no extensive theoretical or semi-empirical solutions available to predict these solutions with reasonable accuracy. The prediction of jet-lift propulsion performance with the myriad of flow turning devices, swivel and cascade nozzles, deflectors, and diverters is almost entirely dependent on full-scale rig tests. Lack of correlation between the few model and semi-empirical estimates to compare with full-scale tests, has precluded the development of design techniques.

Since the development of supersonic VSTOL aircraft will invariably entail an afterburning propulsion system, the technical options of deflecting a high velocity/temperature gas flow must be seriously considered. In addition to the basic research being conducted by the major engine companies in flow deflection, continued programs must address the decreased efficiencies inherent in thrust deflection devices, the distribution of thrust to maintain aircraft trim during a throttle-modulated VTO condition, and the safety aspects associated with the potential of afterburner blowouts. The impact of such systems on the flying qualities and controllability of the aircraft must be considered in the proper integration of the vectoring response rates and their influence on the aircraft. Also, the gas flows significantly influence the ingestion and induced force characteristics of an operational vehicle.

Continuation and expansion of the efforts underway in thrust deflection devices is required. Simulation studies conducted jointly by engine and airframe agencies to determine specific VSTOL engine control and integration requirements are necessary for both lift and lift/cruise engine applications. Tests utilizing full-and part-scale systems are required to determine the dynamic interrelationships of RCS bleed levels, temperature/ingestion sensitivity, throttle modulation and thrust deflection with the overall aircraft. Concurrently, the development of simulation techniques to describe these motions is required for proper design tool implementation. Continued interest in jet-lift as a promising VSTOL candidate demands that integration of the propulsion/flight controls interface be treated precisely.

General

With any VSTOL vehicle, the measure of effectiveness of the lift generation system is the primary and overall assessment of its lifting capability.

The parameters by which that effectiveness is measured vary considerably between generic concepts. To date, there is no one standard term from which the propulsion/lift performance can be compared for all systems. What is meant by performance predictive capability also can take many definitions.

There are three major topics constituting performance: Propulsion Sizing, Environmental Effects, and Lift Generator Performance. Although somewhat different in their scope and considerably different in the methods and research available for their prediction, they are discussed together to expedite a continuous approach to propulsion estimation capability. The evaluation of performance predictive capability includes the past VSTOL propulsion systems as well as some of the lastest techniques and methods.

Thrust and specific fuel consumption (SFC) are not the only direct measures of propulsion performance; the prediction of other engine parameters could equally impact the operational environment of the vehicle. Turbine exhaust gas temperaturn (EGT) is one such measure of propulsion performance. The EGT estimation techniques for the AV-8A and XV-5A underestimate test results which introduces speculation as to potential causes of hot gas ingestion being higher than predicted for both of these aircraft. Comparison of power and thrust coefficients for the X-22A ducted prop system show good correlation between the predicted and measured static performance.

Over the past half century, much attention has been directed toward the prediction of cruise performance for propellers and rotors by analytical and experimental methods. In view of the relative importance of static thrust for VSTOL, one would expect that even greater attention would have been directed toward prediction of static thrust (hover) performance. Surprisingly, with few exceptions, this situation has not prevailed. Initial thrust comparisons of the VZ-2 (Boeing-Vertol tilt wing), show a 11% maximum difference between theory and model/flight test results. Even for the follow-on of the tilt-rotor program, the XV-15, the theoretical estimates of hover performance continue to underestimate the model and full-scale test data by as much as 22%. Tilt-wing performance estimation is considerably better, however, than the tilt-rotor aircraft - CL-84 and XC-142 show close agreement between the theoretical thrust performance with rig or flight test data.

The static performance of an ejector system shows that considerable differences between predictions and measurements can exist. For example, the actual augmentation ratio achieved by the XV-4A in flight testing was 10% lower than predicted. Only by considerable full-scale rig testing and many ejector configuration changes was the predicted value achieved during a rig test. However, even the latest published information on ejector augmentation design shows poor agreement with scaled test data.

The propulsion technology must be evaluated not only for its capability to estimate propulsion system design performance, but also the additional sizing and environmental effects that significantly impact VSTOL vehicle performance. The VSTOL vehicle requires a high power/weight ratio to satisfy its vertical flight mode. High ambient temperatures, hot gas ingestion, propulsion sizing, and aircraft growth all have vital implications for the overall performance of a VSTOL vehicle. It is possible to equip the propulsion system with compensating provisions such as short-term VTO ratings utilizing engine overspeed and/or increased turbine temperatures, water injection, and plenum chamber burning to augment the lifting power of the system. The desirability of establishing specific rating criteria for VTO and reducing the sensitivity to inlet temperatures can be demonstrated. Although several engines have been operated at higher speeds and temperatures for VTO operations, only the P.1127/Harrier engine was designed initially to incorporate both overspeed and water injection to increase the available VTO thrust for short periods of time. Although only limited predicted and actual data are available for this engine, results show that its performance generally can be predicted adequately from thermodynamic and cycle calculations. What is not easily predicted, however, is the effect of such operation on the life and maintainability of the engine.

In the VSTOL aircraft operating environment, the presence of hot exhaust gases for most vehicle concepts raises the specter of engine stall at critical VTO conditions. Several aircraft, the most notable of which are the X-14, XV-5A, VJ-101C, and XV-4B, suffered severely from ingestion-induced stalls to the extent that compromises had to be made in operating the vehicles. It is known that the increased ambient conditions decrease the available surge margin for the compressor, but the actual mechanism of a stall propagation is still unknown. Several flight test and development test results have indicated that not only is high temperature ingestion conducive to stall, but the rate at which the temperature differential is presented to the engine is a factor. To date, no techniques or methods have been proposed to accurately predict this phenomenon but the impact of its criticality should be realized as the VSTOL community considers higher exhaust temperature lift and lift/cruise engines.

There are other facets of performance that must be considered. The major areas are: performance estimation in the presence of vehicle interference (the so-called airframe/propulsion compatibility); the effect of ground proximity on the lift-generator performance; and forward speed effects. The first usually affects to a greater extent the lower disk-loading VSTOL aircraft in hover and low speed because of flow interaction between the aircraft and the lift mechanism (i.e., rotor, propeller). The presence of the ground can have an adverse effect, not only on the vehicle, but on the performance of the propulsion system. For example, the augmentation ratio for ejector systems is sensitive to ground-induced backpressures. Also, ducted propeller performance is effected by ground proximity as evidenced in time-dependent thrust variations. With respect to the forward speed effects, the fan-in-wing/body VSTOL configurations are most sensitive since crosswind flow degrades fan performance.

In general, there are few theoretical and semi-empirical methods available today to provide the predictive capability over the full range of VSTOL operations.

For the prediction of the effects of propulsion sizing, thermodynamic and cycle calculations will provide adequate results; however, the effects on the engine service life are considered inadequate, primarily from lack of experience in this area. As would be expected, full-scale rig testing can give sufficiently accurate data to formulate flight results by controlled testing conditions, including simulation of the actual VSTOL usage through cyclic testing. Somewhat analogous to the sizing prediction techniques, the environmental prediction methods can produce adequate results with regard to ambient temperature effects. The problem area lies in the ability to predict the effect of transient temperature distortions on the engine or lift-generator performance. It should be noted that model application to these problems is not considered an efficacious method for predicting final performance because it does not accurately simulate an engine system's operating characteristics.

Prediction of the lift-generator performance varies somewhat depending on the generic concept. The major drawback to adequate theory/semi-empirical prediction centers about the proper definition of the aircraft/propulsion interactions and the installed characteristics, including ground proximity and forward flight effects. Model tests provide a better prediction capability but have been limited by the inability to simulate the propulsion system flows properly. The scaling laws necessary for extrapolation to the full-scale vehicle have not been firmly established, although some work is being conducted by several agencies to determine flow and performance parameter correlations. Again, the rig testing permits good prediction of performance particularly for the static condition; however, unless the total vehicle is simulated in these rig tests, they have limited value in predicting transition performance because of the high interaction between the propulsion and airframe flows.

Direct in-flight measurement of thrust is not possible for any of the generic types of VSTOL, but methods have been developed to compute thrusts based on measured engine/propeller/fan speed (RPM) and other measurable quantities. These methods are dependent on empirical data from rig testing to account for installation, transmission, bleed, and other losses and do not always account for mutual interference effects of multiple lifting units or airframe/lifting unit interaction effects. That is, while it is possible to establish the combined airplane/engine hover performance (e.g., hover weight and fuel consumption), it is not possible to measure lift (thrust) losses caused by interference effects; to do this, the thrust must be computed utilizing flight measured parameters (such as EGT, engine or propeller RPM) in conjunction with thrust stand measurements of either installed or uninstalled engine thrust.

Although the Air Force has done some preliminary investigations into the feasibility of measuring both the thrust magnitude and direction through the use of piezoelectric strain gauges on the engine mounts, the system was never developed to a practical application. While the lack of thrust measurement capability does not seriously affect VSTOL operations, an indication of thrust (or thrust margin) available would be an aid to the pilot, and would give the designer confidence in the sometimes complicated thrust estimation procedures available for the various generic types.

7. Technology Area - System Losses

- nozzles
- gear/transmission losses
- thrust deflection
- end wall losses

Estimation Capability

- theoretical/semi-empirical: poor
- empirical
  model: poor
  rig : good - poor

Measurement Capability

- flight test: **poor**

Remarks

- theory/semi-empirical
    - poor for deflectors, diverters, swivel nozzles, flow turning devices
    - good for straight nozzles, ducting, gear/mixing box losses

- empirical-model
    - flow scaling and area/parameter scaling problem
    - flow leakage
    - forcing
    - frequency difference

- empirical-rig
    - good for higher disk loading vehicles
    - full scale duplication possible
    - aircraft interference effects on flow

- flight test
    - difficulties in obtaining appropriate measurements with flight-worthy instrumentation lead to heavy reliance on rig testing to establish losses

Technology Summary

The performance of mechanisms associated with generating a vertical lift component for a VSTOL aircraft impacts system performance considerably. The mechanical and aerodynamic vectoring schemes used to turn the propulsive flow - transmission and mixing box systems, hot gas ducting and diverters, and diffusion ducts - all induce losses on the lifting efficiency of the vehicle. In many instances, these losses are heavily dependent upon the aircraft attitude. The prediction of these losses to estimate flight performance throughout the VSTOL regime is a formidable task.

VSTOL deflection of exhaust gases creates static pressure gradients within the deflector system that propagate upstream and produce an unsymmetric backpressure on a turbofan or fan-powered device. As the uneven distribution of backpressures is applied to the fan, the flow is cyclically accelerated and decelerated. This condition can cause stall/surge of the fan and induce a significant lift loss. The effects of this backpressure on both the dynamic and steady-state performance is unknown.

Likewise, the pressure losses associated with the vectoring nozzles and louvers, used primarily for jet-lift and fan-in-wing aircraft, introduce many prediction problems. The use of an afterburner and the vector angle influence vehicle performance considerably. As much as 3 percent T/W margin could be sacrificed by deflecting an afterburning system. Methods for predicting combustion efficiency do not provide detailed guidance for designing a vectorable afterburning nozzle system.

The techniques for predicting nozzle thrust coefficients for various exit and duct turning angles are highly empirical in nature.

The performance of flow diverter doors on the pitch fan of the XV-5A has a very pronounced impact on the control sensitivity of the aircraft. Model test results and predicted performance show a close correlation over almost the entire range of control deflections.

The duct losses inherent in diffuser ducts of augmenters and ducted prop aircraft can represent a significant portion of actual performance derived from such a system. Prediction techniques in this area require considerable effort to upgrade the capability.

There are, in general, four primary loss mechanisms that degrade incompressible ejector performance: inlet loss; momentum coefficient of the primary stream; thrust efficiency of the primary nozzle; and friction, mixing, and diffuser losses. Taken individually, each significantly reduces the augmentation ratio. When these losses are combined, the effect can be even more drastic. Individual losses and their interaction with each other still remain outside the realm of accurate prediction.

Loss predictions from the existing data base are expected to be poor for deflectors, diverters, diffusers, swivel nozzles, hot gas ducting, and mechanical losses. Model application to prediction of losses also is considered to be poor for all the generic vehicles primarily because of the scaling problems connected with both flow and size. The rig tests are considered to provide the best prediction capability particularly for the higher disk loading vehicles. For the lower disk loading vehicles and those vehicles with lift-generation systems sensitive to ground proximity and airframe interference, only a full-scale vehicle test is expected to provide adequate prediction capability.

Free-flight determination of system losses becomes a problem because of the difficulty often encountered in measuring appropriate parameters, in locating transducers so they will not have an adverse effect on aircraft operation or on the quantity to be measured, and in routing wiring bundles within the geometry of the airplane structure. For these and other reasons in-flight measurement of systems losses is often not pursued unless they present some threat to operation of the aircraft. The tendency instead is to utilize rig, tiedown, or other estimated data as the basis for correcting measured data in the determination of total aircraft performance.

8. Technology Area - Ground Environment

- temperature of ground flow
- pressure of ground flow
- velocity of ground flows
- flow interaction
- fountain formation & characteristics

Estimation Capability

- theoretical/semi-empirical: poor - none
- empirical
    model: poor
    rig : poor - good

Measurement Capability

- flight test: good

Remarks

- theory/semi-empirical
    - methods OK for pressure profile
    - poor for flow interaction, fountain characteristics, mutliple jets, unequal strength jets, wind effects
    - temperature prediction worst
    - height/pressure/temperature scaling
    - nozzle mixing characteristics
- empirical-model
    - small and large scaling effects
    - flow and temperature scaling problems
    - induced flow simulation necessary
- empirical-rig
    - good results
    - no scaling effects

Technology Summary

The design of a VSTOL vehicle and its operational deployment characteristics is highly dependent on the environment imposed by the downwash flow from its lifting mechanisms. Flow complexity and the lack of adequate prediction techniques to analyze these flows in the ground plane exist. The temperature/velocity/pressure profiles within the flow regions can be estimated reasonably only for limited and generally simple configurations. Since basic understanding of ground flows and their interactions is essential to the formation of methodologies for hot gas ingestion and induced force disciplines, a systematic program with the ground flow as a baseline is required. In view of the poor coorelation of data, the ground separation parameters including wind effects must be explored further.

A parallel effort of theoretical/semi-empirical and model test approaches is necessary to determine adequate design estimation techniques. The approach should consider the treatment of turbulent mixing of jet flows particularly as applicable to the free and ground jets. A detailed understanding of the induced flow field generated by these flows in the presence of wing/body constraints must be sought. Flow characteristics of the fountain and flow interaction zones should be determined with special attention directed toward accurate modeling of the upward flow momentum and temperatures and stagnation line characteristics.

The model tests should concentrate on the determination of the flow path in the presence of an aircraft and its operational environment. Specific emphasis should be directed toward multiple jet configurations and unequal strength jets to simulate typical lift and lift/cruise configurations. The trajectories of the near- and far-field flows and their flow properties are necessary, including fountain direction and momentum and wind effects on all flow paths. Azimuthal variations in flow characteristics should be determined for various combinations of jet inclinations and dynamic aircraft attitudes to determine potential influences to the aircraft during IGE (In Ground Effect) maneuvers. Of major importance is the determination of model scaling parameters as well as height/spacing/temperature/pressure scaling laws for jet simulation. Model tests should be conducted, in great part, in conjunction with hot gas ingestion/induced forces tests. A combination of wind tunnel (for wind effects) and special propulsive model rigs should be utilized. Various scale-model tests of the same configuration and comparison with similar full-scale tests (e.g., AV-8A, VAK-191B) are required to provide verification of scaling laws.

Implicit in the search for ground profile descriptions is the influence of the flow on the operational aspects of VSTOL aircraft. Ground erosion/foreign object damage/deck heating prediction techniques should be investigated and verified by full-scale velocity/temperature flow tests. Ground personnel and equipment safety in the presence of the ground flows should be considered. Crew systems requirements and deployment constraints should be identified to provide a safe and realistic preview of operational implications.

General

One of the most serious operational problems associated with VSTOL aircraft stems from the downwash flow from its propulsion and lift generation system. Downwash can impose the most severe design constraints on a vehicle. The three major topics to be discussed are shown in Figure IV-8-1. Every VSTOL vehicle manifests, to some degree, a propulsion-induced effect on the aircraft or the surrounding environment. Overcoming this effect requires considerable sacrifice to the vehicle goals. These phenomena are usually generated by the high-temperature, high-velocity flows from the lift generator impinging on the

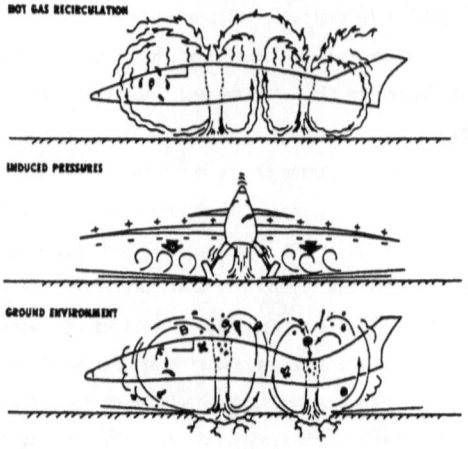

Fig. IV-8-1. Ground Proximity Propulsion Induced Effects

ground, spreading radially in a thin turbulent sheet until deviated by an obstacle or interaction between flows, and finally rising to have some interaction with the aircraft. To assist the reader in a basic understanding of the scope of these problem areas, a brief outline of the symbology and definition of the flow patterns for each phenomenon are provided.

The ground environment reflects the pressure, temperature, and velocity fields included by the propulsion flow at (or slightly above) the ground plane. As shown in Figure IV-8-2, the flow is separated into two regions of concern: the near field, which is usually defined as the area within the confines of the aircraft platform, and the far field, which constitutes all area outside the near field. The radially flowing jets form lines of interaction (stagnation lines) at locations where the dynamic pressures of the flows are equivalent and combine to produce regions of upwash commonly called fountains. In the far field, the wall jet expands outwardly until the flow velocities are reduced sufficiently that the buoyancy effect of the warmer air causes the flow to rise or, in the case of a wind, is separated at a far-field stagnation point. The major concerns with the ground environment are the direct influence on aircraft operational restrictions resulting from structural heating, ground erosion/surface preparations, and ground crew and equipment compatibility. The primary governing parameters affecting the ground flow are the disk loading of the vehicle (i.e., downwash dynamic pressure) and the height, attitude, and flow conditions of the lift generators. As would be expected, higher disk loading vehicles impose higher

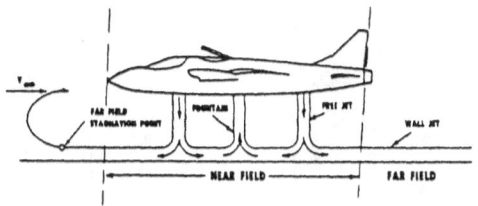

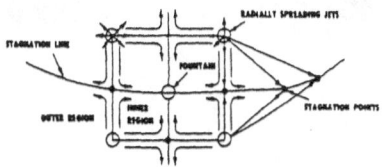

Fig. IV-8-2. Ground Footprint Symbology and Region Definition

velocity and temperatures on the ground environment. Typical velocities and temperatures in the vicinity of a modern lift engine are shown in Figure IV-8-3.

The estimation capability of the flow conditions within the ground environment varies considerably for each generic type. Most experimental tests and theory have been directed toward the jet-lift and fan applications, primarily because of the high flow conditions and their potential impact on the aircraft. The decay of dynamic pressure and velocity has been found to be predictable to a much greater degree than ground temperatures. For single jets and double jets of equal strength, good results have been obtained in estimating the pressure field at various inclinations of the jets. The pressure estimation technique is supported by data from several models and correlates well with AV-8A flight test measurements.

Although pressure decay can be predicted accurately, temperature decay cannot be. The dependency of ground temperature on the number of jets is very marked and correlation of AV-8A test data is poor. The prediction of the absolute magnitude of temperatures remains a major problem area.

A modified theoretical approach based on model data was used by Canadair to predict the ground flow velocities for the CL-84, and the results have been compared to flight test data. In general, the prediction method overestimates the velocities.

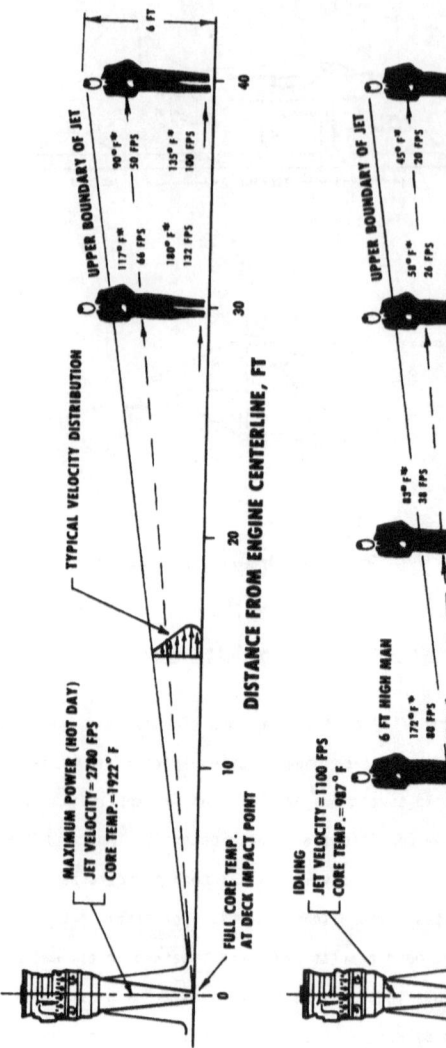

Fig. IV-8-3. Velocities and Temperatures in the Jet Footprint of XJ99 Lift Engine

Crosswinds not only affect the total pressure and temperature distributions within the ground jet but also the radius of separation of the flow from the ground. Tests with a moving rig indicate that the flow separation for a single jet occurs when the momentum of the wall jet approaches that of the crosswind. Applying this similar rationale to two-jet small-scale models shows that the prediction technique does not provide good coorelation with test data. Only a fair correlation is obtained at the lower wind velocities with large discrepancies evident at the higher wind conditions. Other comparisons indicate that no techniques are available for adequate prediction of crosswind effects on the ground flow, which is unfortunate since accurate determination of the separation radius could be instrumental in determining a theoretical model for far-field temperaturn ingestion by the propulsion system.

A further major concern within the ground flow is the fountain flow. It significantly impacts the aircraft since the upward flow is transported to a region where the hot gases may be ingested by the engine intakes (near-field reingestion), and the momentum of the flow can induce an appreciable force on the aircraft structure. The turbulent nature of the fountain renders a theoretical treatment of the flow very difficult, and only extremely limited small-scale experiments have been conducted to obtain an understanding of the fountain parameters. Tests involving the formation of a fountain by the merging of two two-dimensional cold ground jets showed that the fountain velocity decayed in a manner similar to a free-jet and that the fountain location and direction were directly related to the jet momentum ratios. It should be noted that no tests for three-dimensional flows have been conducted and that application of these test results to a multi-jet aircraft configuration could lead to erroneous conclusions.

For the available theoretical and semi-empirical data base, the prediction methods are considered adequate to estimate the ground flow pressures and velocities up through dual flow sources. The prediction of the flow characteristics in the flow interaction regions and the fountain and the effects of multiple jets, unequal strength jets, aircraft attitude, and wind are considered poor. Of all the flow characteristics, temperature prediction in the ground flow is the least dependable.

Model tests upgrade the predictive capability for the ground flow considerably; however, until adequate flow scaling of the jets and the extrapolation of effects of wind and aircraft attitude for scale models are shown to be applicable to full-scale test results, some apprehension for direct usage exists. Although relatively few flight test data have been acquired to define the ground environment generated by a hovering VSTOL, it is more a matter of test priority than technical difficulty. Unlike most propulsion-induced effects, measurement of downwash (the radial flow along the ground of the efflux from the lifting units and entrained ambient air) is simplified by the fact that the instrumentation and its associated wiring and recording devices do not have to be installed within the geometric constraints and thermal or vibration environment of the aircraft. Despite this fact, a paucity of quantitative data has been obtained in flight because of one or more of the following reasons:

- Limited hover time available from fuel considerations
- Difficulty in holding precise spot hover
- Possibility of flow distortion by existing wind conditions
- Lack of facilities to simultaneously measure and record the desired data at diverse locations to assure that data were for the same flight and wind conditions
- Priority of other test requirements.

Although some cursory measurements of velocities and temperatures were made on the XV-5A by Ryan and the AV-8A by the Naval Air Test Center, the most complete data from flight were taken by Canadair on the CL-84-1. In these tests, groundwash data for a single-rotor helicopter of comparable weight were also gathered for comparison. In most cases, only a qualitative assessment of the flow field has been conducted (e.g., dust cloud behavior during desert landings of the XC-142 and XV-5A, far-field reingestion phenomena for various jet-lift vehicles, aircraft disturbances as the VAK-191 encountered its own ground disturbance in transition).

To be useful for comparison of the ground flow characteristics of different vehicles either within or across generic types, agreements on, and adherence to, a standard measurement technique are desirable.

9. <u>Technology Area - Hot Gas Ingestion</u>

- ingestion of exhaust gas flows

<u>Estimation Capability</u>

- theoretical/semi-empirical: <u>poor - none</u>
- empirical
  model: <u>poor</u>
  rig : <u>poor - good</u>

Measurement Capability

- flight test: good

Remarks

- theory/semi-empirical
  - single flow prediction poor at best
  - multiple flows very configuration oriented
  - no treatment of viscous mixing/boundary layer interaction
- empirical-model
  - flow scaling parameters needed
  - dynamic effects
  - aircraft attitude
  - wind effects
  - jet modeling
- empirical-rig
  - good only for full-scale vehicle
  - no transition effects
  - poor for isolated rig (not representative of aircraft)

Technology Summary

Estimation techniques based on both theoretical and scale model tests are inadequate to describe over-temperature effects caused by recirculation of hot exhaust gases. Although some model tests show that specific scaling laws provide adequate estimation of full-scale tests (e.g., P.1127), similar correlations on other tests (DO-31) employ an almost completely opposing scaling law rationale. Consideration must be given to the potentially high inlet over-temperatures possible from future VSTOL engines and the accompanying thrust loss and impact on operational safety. Of equal importance are the configurational influences, wind and ground roll velocity effects, and engine temperature/stall characteristics.

As discussed in the subsection on ground environment, an accurate modeling of the ground flow and near-and far-field upwash flows will provide a solid base from which to continue a theoretical approach to simulate the gas path trajectories back to a specific configuration's inlet. Until such accurate flow modeling techniques are available, it is necessary that emphasis be concentrated on developing accurate model scaling techniques and seeking correlations from specialized model tests to provide design guidance.

A comprehensive and systematic program is required to address the establishment of relationships between vehicle configuration, inlet location, jet number and location, and aircraft height as well as the flow properties of the exhaust gases. A flexible configuration test rig that will permit a wide variation of configurations can be used. Detailed instrumentation of the model configuration, inlet system and the flow field as well as flow visualization techniques should be employed to map the near- and far-field flow paths accurately. Space-time thermal mapping of the inlet system should be incorporated to provide meaningful, standardized ingestion indices that would provide quantitative information on the spatial and temporal variations of the inlet temperature characteristics, including localized thermal distortions and temperature transients (those parameters judged to be the major influences of engine stall).

The effects of wind and ground roll velocities on the ingestion characteristics of VSTOL configurations must be identified. In addition to wind tunnel tests, open-air rotary circulation rigs can provide satisfactory test results on the limited configurations tested. Determination of the influence of inlet location, inlet size, and jet exit conditions on ingestion characteristics at low forward speed IGE will provide insight into the operational alternatives to reducing hot gas effects. Along similar lines, mechanical "fixes" to the aircraft to reduce the gas flow ingestion are possible. Several potential methods have been identified, including blowing slots in the fuselage to divert the near-field gas flows from the inlet location and the use of landing gear and engine doors to form blocking surfaces for the gas flows. Surface fixes should be investigated despite several aircraft being operationally compromised by restricting VSTOL flight from grid surfaces. Alternate schemes should be examined whereby external gas deflectors on the deck or surface could channel the flow to regions that produce less effect on the inlet temperature rise. Within the propulsion system components themselves, nozzle configurations that promote high jet mixing could provide significant strides in reduction of ingestion effects. Investigation of such devices, however, must address the compromises in lifting capability both from the standpoint of expected lower nozzle thrust coefficients as well as decreased aerodynamic lift from fountain impingement.

Verification of the techniques generated by such studies should be conducted, where possible, by comparison with test data obtained from past and present VSTOL vehicles. Utilization of the AV-8A, X-14, and XV-5B (if possible) as testbeds for full-scale correlation should be considered.

General

During the low-speed and VSTOL flight modes, the exhaust gases from the propulsion system tend to be reingested by the engines because of the ground flow properties described earlier. Since the output of engines is sensitive to intake temperature and the exhaust gases in some concepts are extremely hot, the ingestion of only a small proportion of the exhaust gases can result in significant thrust losses or even engine stall. Jet-lift vehicles suffer from hot gas ingestion to a much greater extent than the lower disk loading vehicles with the possible exception of fan-in-wing. In fact, none of the propeller/rotor or augmenter aircraft show any effects caused by hot gas ingestion. However, because of the increased exhaust gas temperatures of future propulsion systems and the major influence of configuration in the dependence on hot gas ingestion, this problem should be addressed by the designer. It is known that the mean excess intake temperatures can be as much as 10 percent of the engine nozzle temperatures depending on inlet arrangements. Thrust losses of about 1 percent of every $3^{o}F$ of inlet temperature rise have been reported for the X-14 and XV-4B. With the higher cycle engines currently proposed for VSTOL, this loss rate could be as much as 1- to 1.5-percent/$2^{o}F$. Thus, the importance of minimizing reingestion cannot be overemphasized.

The major influencing factors governing hot gas ingestion, in addition to the nozzle temperature, are aircraft height, forward speed, and configuration. Since reingestion is an effect caused by the ground influence on the flow, it will decrease rapidly with aircraft height. At takeoff, the fountain (near-field) reingestion is the primary temperature influence to about 10 feet above wheel height, at which time the far-field ingestion predominates.

The effect of forward speed is to progressively reduce the near-field ingestion while increasing the far-field ingestion. At a critical forward speed, the ingestion reaches a maximum. Additional speed blows the exhaust backwards beneath the intakes, and ingestion rapidly decreases. The configuration of the powerplant and inlet system has a pronounced effect on this speed: nozzle canting for the VAK-191B; high inlets for the VJ-101C; and low, large inlets for Harrier. Aircraft layout can be a key factor in minimizing ingestion from the near field. In general, the longer the path between the fountain and the inlets, the greater the opportunity for the hot flow to mix or be deflected from the inlets.

Although there is a qualitative understanding of the hot gas ingestion mechanism, no adequate systematic, comprehensive theoretical or semi-empirical approach has been developed and verified by full- or large-scale tests. Several investigators have attempted to provide a fundamental guideline for estimation of the gas flow, but results have fallen far short of a suitable prediction technique. The poor results arise mainly from the inadequacy of predicting the ground flow conditions and, more predominantly, from the uncertain flow path of the hot gases in the near field. Several partial semi-empirical techniques are available, the most notable of which is the one that combined results of many tests on different parts of the hot gas flow to provide a single technique. The method falls short primarily in its application since it does not directly tackle the near-field ingestion and is limited to equal strength nozzles. The method provides for variations in nozzle pressure ratio, temperature, diameter, ground height inlet location, and crosswinds for VSTOL flight only. The technique has not been verified by comparison with test data.

A theoretical method has been described that is based on potential flow theory for determining the increase in inlet temperature from a single isolated jet discharging onto a flat surface. Although it considers only a very simple configuration, the results are validated by experimental data; therefore, it is a very useful tool for studying the significance of specific flow characteristics. The main parameters of the study are wind, inlet location, jet temperature, and nozzle exit velocity (Mach number). Several other theoretical or semi-empirical prediction methods are available but are considered either too simplified to warrant general usage or are completely unsubstantiated by test data.

The use of scale models to predict hot gas ingestion has drawn the most attention as the primary tool for analysis. Considering the complexity of the flow field to be analyzed and the inadequacy of the fundamental characteristics of dependency, it probably is the most cost-effective technique. Basic research has been conducted into the scaling laws applicable for model testing for hot gas ingestion by use of a simple two-jet/sucking inlet arrangement. Extrapolations have been made to limited P.1127 data, and a fair correlation was achieved. However, the important results were scaling laws for model tests. Subsequent tests by Rolls Royce showed that the scaling laws provide a correlation technique that, while not perfect, can be applied for preliminary estimates. Although the method uses experiemtnal data, wind effects data have not been demonstrated; thus, some questions remain as to its validity under all operating conditions.

Although the scaling laws and details of the tests are unavailable, some comparisons of predicted hot gas ingestion temperatures from scale models with full-scale flight tests on various aircraft are available. There is close correlation between model and full-scale temperatures for the lift engine and main engine intakes of the VAK-191B. For the DO-31, the correlations are mixed. With a scaling technique based on a direct comparison, i.e., identical nozzle Mach numbers and temperatures, there is good correlation between the full-scale testbed and a 1/20-scale model. However, when other scaling parameters were attempted, the correlations between the full-scale testbed and the models are extremely poor. Only direct scaling provided good comparison.

Although no information is available to describe the methods used for prediction of the expected hot gas ingestion, estimates were made prior to testing for the XV-5A and VJ-101C-X1 (nonafterburning). Comparisons of the estimates with flight test data show that prediction techniques for hot gas ingestion could be improved.

From the previous discussions, inadequacies and inconsistencies exist for estimating hot gas ingestion. Theoretical and semi-empirical methods offer no general solution for consistent and confident "ballpark" estimates. To a much lesser extent, estimation confidence from model testing is considered poor and will continue to be unless adequate scaling parameters are proved or determined. The rig tests naturally provide the highest confidence for estimating the flight characteristics. Care must be taken, however, that sufficient aircraft geometry is included to assure that flow paths are simulated.

There have been no attempts to map the three-dimensional temperature field created by actual operations of vehicles, but measurement of engine-inlet temperature rise from far-field reingestion is easily accomplished. Near-field reingestion may not be evenly distributed across the inlet, requiring more extensive instrumentation if inlet temperature profiles are required.

10. Technology Area - Induced Forces & Moments

- propulsion induced lift loss/gain (IGE & OGE)
- induced moments from exhaust flow
- induced moments from intake flows
- induced moments from downwash reflection

Estimation Capability

- theoretical/semi-empirical: poor - none
- empirical
   model: poor
   rig : poor - good

Measurement Capability
- flight test: poor

Remarks
- theory/semi-empirical
  - flow interactions uncertain
  - fountain characteristics
  - induced flow
  - configuration parameters
  - wind effects
  - efflux/influx characteristics
- empirical-model
  - model and configuration dependency
  - flow scaling
  - dynamic flow interactions
  - jet flow
- empirical-rig
  - same as hot gas ingestion (HGI)
- flight test
  - self-induced forces and moments can be inferred from vehicle response, but not measured per se

Technology Summary

The aerodynamic forces induced by the lifting system are most significant. Every VSTOL vehicle manifests to some degree the significance of this flow field effect. The unique feature of these induced force and moment phenomena is that they are equally significant both IGE (in ground effect) and OGE (out of ground effect) and from the hover through transition velocity regimes for most concepts. Although considerable test efforts here and abroad have been initiated and some limited theoretical treatment has been attempted, the results afford a poor estimation capability as applied to new design concepts. Because of the magnitude of induced forces and the severe consequences that they impose on stability and control as well as the initial design of the aircraft, this area must be emphasized.

Within the theoretical approach, continuation of flow modeling techniques are required. The technical methodology for the OGE influences must be expanded to include IGE flow models and other potential flow aspects (e.g., efflux flow) based on the verification tests being conducted by NASA/Langley.

The semi-empirical approach offers a potential for adequate induced-lift loss estimation. The limitations and constraints of this methodology should be identified by validation with flight test and full-scale model test results. Some independent research and development efforts on scale models have shown promising trends for a hovering vehicle IGE and have provided indications of the relative importance of various induced-lift loss parameters.

A single, continuing test effort should be undertaken utilizing the results of previous programs to determine in detail the parametric effects of configuration and propulsion variables. A flexible-configuration test rig is necessary to permit testing of a wide range of configurations. The major emphasis areas for model testing and design guidance should be wing planform and height, jet location and spacing, flow scaling parameters (including nozzle pressure ratio and temperature), and segregation of the lift components (e.g., fountain lift, wing/fuselage suckdown). Although methods of measuring this latter component involves complex balance and sting arrangements, the definition of the relative contribution of each flow component will provide significant insight into design estimation techniques. Different scale sizes of the same model should be tested to determine scale effects, if any. Determination of the effects of aircraft attitude and maneuvering on the induced forces and moments experienced by the model, as well as flow scaling laws for simulation of inlet flows is primary for assessing the dynamic interplay between the control and lift systems.

General

The lift generators of a VSTOL aircraft can impose gross constraints on the mainstream flow over the aircraft surface during both hover and transition flight. As shown in Figure IV-10-1, the interference effects on the inlet, wing, fuselage, and tail caused by induced flow generated by the influx/efflux of the lift system can influence both the lift and stability of the aircraft. These induced effects are present regardless of aircraft height (in or out of ground effect) although the IGE forces generally predominate. The aerodynamic interference effects during the low speed and transition phase can be of the same magnitude as the aerodynamic loads; therefore, intricate knowledge and understanding of their influencing parameters is necessary for adequate control design. The induced flow and moment effects are not limited to the high disk loading vehicles as was the case in the hot gas ingestion, but are for the full range of generic vehicles. Typical qualitative lift effects are shown in Figure IV-10-2. Most of the theoretical and testing work, however, has been concentrated in the jet-lift and fan areas, since the greatest "pumping" action of the flow is experienced with these concepts.

There are several theoretical and semi-empirical treatments for the behavior of jet flows on an accompanying surface. For example, a simple working formula has been described based on curve fitting of model data to estimate the lift loss of a jet/wing combination. In extrapolating these results for

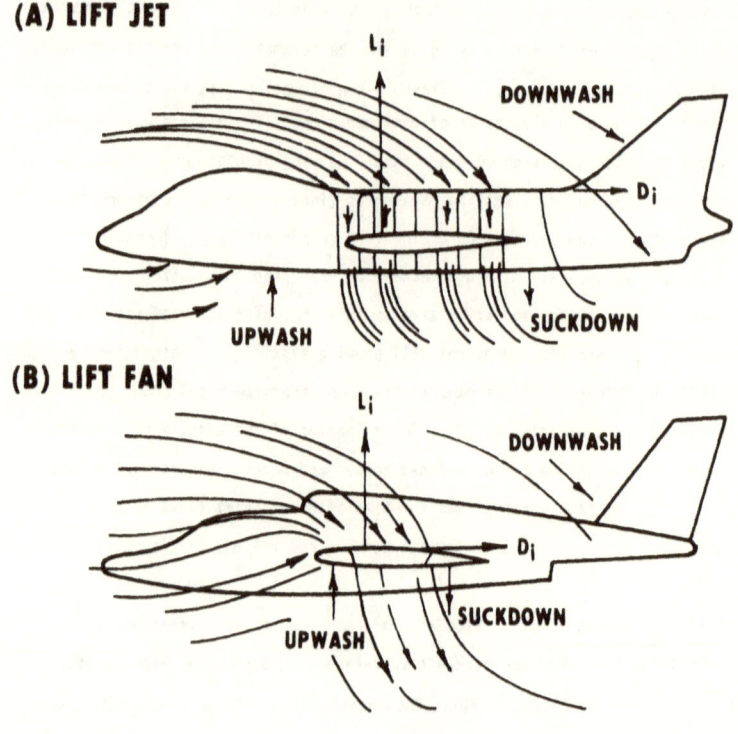

Fig. IV-10-1. Lift Jet and Lift Fan Interference Forces in Transition

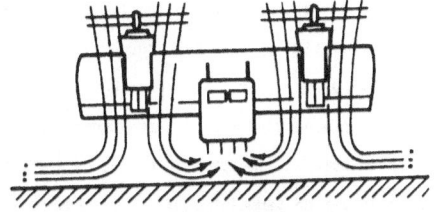

FAVORABLE GROUND EFFECT

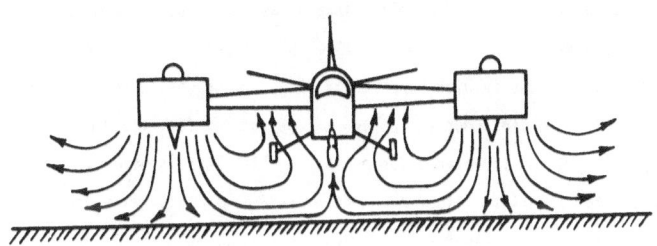

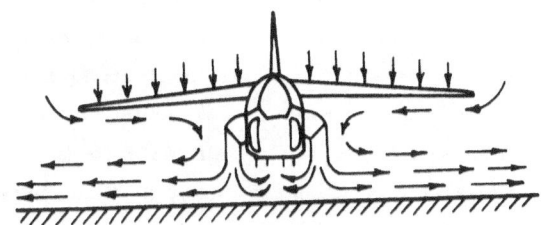

NEGLIGIBLE GROUND EFFECT

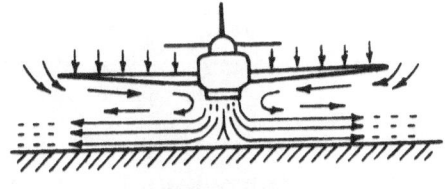

UNFAVORABLE GROUND EFFECT

Fig. IV-10-2. Typical Lift Loss Effects on Aircraft

full-scale comparison, no corrections are applied for scale effects; yet, good correlation between a full-scale jet exiting through a flat plate and a X-14 model and full-scale data has been obtained. One should not be so naive as to think that this method is a panacea for lift loss prediction, since it has predicted well only for clustered, jet configurations on low-wing aircraft and plates. The lack of correlation techniques, it is further pointed out, is evident when considerations for wing heights, exit geometries, efflux velocity distribution, flow decay rates, and aircraft attitude are concerned. For the forward flight case, a "vortex sheet" theory was expounded, but its application is limited only to far-field effects since it assumes that the momentum in the jet flow remains constant and the jet flow is parallel to the freestream direction. Comparison of the theory with experimental data showed considerable differences.

The vortex and sink-doublet representation of the flow field has been developed to provide techniques for estimating lift loss in hover and transition (OGE only) for jet-lift, lift-fan and deflected-thrust VSTOL aircraft. In general, the three-dimensional potential flow model is well founded and gives relatively good agreement with single jets. However, significant variations between the theoretical estimates and wind tunnel results are evident for multiple jets.

Several techniques for estimating influx flow effects and fuselage lift losses are available. The inlet model, although seemingly inclusive of all effects, does not compare well against 1/6-scale XV-5A force and moment data. It has been suggested that the addition of fan exit effects to the empirical model could improve correlation. Theoretical estimates for fuselage forces and moments likewise fall short of acceptable correlation with model test data. Despite the relatively poor capability of the methods as design tools, the techniques are sound in principle and the trends of the data can be estimated. NASA/Langley is conducting part- and full-scale tests to validate these efforts which should provide better insights as to their predictive capability.

Another purely theoretical approach has been made to estimate induced losses during hover flight. It includes interesting data comparisons from tests to examine various influencing effects of the aircraft and jets. Considerable detail is expended in the development of a theory for prediction of IGE lift losses and the application of this theory to a circular disk planform yields good agreement with experimental results.

Only limited theoretical/semi-empirical prediction methods have been uncovered for the other generic type of aircraft. One such method is a simplified inviscid two-dimensional flow model that provides ground proximity effects on a shrouded prop configuration by energy considerations. The theory correlates well with test data.

An investigation has been made to determine the force and moment contribution of jet-lift and lift-fan inlets in transition based on momentum theory and potential flow considerations. Comparison of the results of this method with part-scale model tests for the jet-lift inlet and with full-scale fan-in-fuselage and XV-5A test results indicates that the method predicts poorly. The large discrepancies in lift and moment predictions may be due to pressurization effects on surfaces adjoining the flow exit which were not considered in the model.

The model tests conducted for various aircraft provide, for the most part, adequate estimation of the ground or flight test induced forces. The most notable example of lift loss is for the X-14. The X-14 with the short landing gear was experiencing lift loss on the order of 40% at wheel height. A scale model was used to investigate the problem and provided a good data baseline from which to verify the final result.

Although measurements of induced forces and moments are not possible from free flight, they can be inferred from measurement of aircraft response. Wind tunnel tests are considered more reliable, because in the flight tests the interference is determined as a residual term in the force and moment balance, whereas it is measured directly in the wind tunnel. Despite the fact that rigs or test stands tend to distort the flow field producing these effects, an estimate from these sources is generally preferable to free flight measurement because of the difficulty in controlling and/or defining the flight environment and the inherent dynamic nature of the aircraft in flight.

The theoretical/semi-empirical estimation capability is poor at best. Although several techniques have promise for estimating the induced forces and moments, none exists that could provide the required accuracy for design prediction. The major areas of uncertainty lie in the flow interaction and interference with both the ground and the aircraft, and the proper description of the influx/efflux. Prediction of wind and ground flow effects reduces the opportunity to advance the capability.

Model testing also provides only poor estimating capability in this area. The scaling laws, particularly for flow and area simulation, still must be proven; suitable techniques must be made available to simulate the jet flow and aircraft interference effects. Trends indicate that these scaling laws could lie within potential or nonlinear aerodynamic theory.

Rig testing remains the best of the predictive devices. The comments provided for the rig tests under the hot gas ingestion summary also apply to this discipline.

STABILITY AND CONTROL

INTRODUCTION

This section presents the results of examining selected stability and control qualities to provide one of several major inputs toward assessing overall VSTOL technology. The characteristics discussed are:

- Equilibrium Trim
- Static Stability
- Control Power
- Gust Sensitivity
- Coupling
- Dynamic Stability
- Height Control

It is not possible to make positive determinations of design estimation capabilities relevant to VSTOL stability and control. Because of a serious lack of flight test measurement data, there is no absolute way to determine, for all aircraft, the credibility of a given estimated quantity, be it derived from theory, semi-empirical formula, or wind tunnel tests.

11. Technology Area - Equilibrium Trim

   Estimation Capability

   - theoretical/semi-empirical
     IGE: none
     OGE: poor - none

   - empirical-model
     IGE: none-poor
     OGE: poor-good

   - empirical-rig
     IGE: poor
     OGE: good

   Measurement Capability

   - flight test
     IGE: good - poor
     OGE: good

Remarks

- theoretical methods are virtually unavailable
- some semi-empirical methods have been successful OGE; nothing IGE
- tilt-wing and tilt-prop/rotor estimates are better since they have been derived from years of helicopter technology
- correlation data are scattered and little are available
- flight-test measurements at low speeds are limited by capability to measure airspeed, sideslip, etc.; IGE measurements depend on ability to operate the aircraft steady state in proximity to the ground

Technology Summary

Flight experiences by VSTOL pilots consistently reflect the annoyance and danger they feel when an aircraft must assume extreme angles of pitch and roll to trim the aircraft in a given state; that is, hovering over a spot and aligning along a given heading in a steady wind of 30 knots or more from any direction. This condition creates an unnatural kinesthetic feel for the pilot. The pilot's external view can become severely distorted in perspective and in many cases completely obscured. Thus, visual cues that are so important in proximity to the deck become either unavailable or erroneously sensed. In addition, such extreme attitudes increase the probability of a deck strike in the presence of unpredictable near-field gusts.

The attitude required to trim a VSTOL aircraft is related to the thrust component available in the wind direction to offset the drag seen by the vehicle, and to provide enough of a counteracting moment to arrest an angular acceleration caused by an imbalance of aerodynamic moments acting on the aircraft. The drag is a direct function of the overall aircraft geometry and planform and is not easily reduced; therefore, it will have a serious effect on development past a major external design freeze point. To provide static restoring moments demands a thrust reserve in the lift generators beyond that required to maintain a minimum acceptable thrust-to-weight ratio and height maneuvering control. This implication is usually very severe since the lift generation system is chosen early in the development cycle and has limited growth potential. The implication, then, of allowing a poor estimate of attitude-to-trim to extend too far into the development cycle is considered serious since it is virtually impossible to overcome in a practicable manner.

Excess attitudes-to-trim would be reflected by a number of operationally oriented factors; for example, pilot training, compatibility with existing visual landing aids, deck crew safety, carrier steaming directions. Implications of more direct importance are those that would incur ship operating constraints, flight restrictions imposed by wind-over-deck considerations, and safety to both pilot and deck crew. Limited visibility (a direct consequence of excessive attitude-to-trim) would always incur a severe penalty on boarding rates.

The main aerodynamic terms required to estimate trim are static forces and moments as well as control effectiveness about and along each axis, including for VSTOL aircraft in particular, the thrust control effectiveness. Few theoretical/semi-empirical methods exist for reliably estimating the basic aerodynamics of VSTOL aircraft especially IGE. There are no techniques for predicting the basic aerodynamics to better than a 30-percent error IGE. Some recent work has provided the beginnings of a promising estimation method for jet-lift and possible fan-in-wing/body aircraft OGE, but it is far from being considered as an "off-the-shelf" reliable method. Attempts to validate some of these methods have been made by wind tunnel testing of model components (e.g., the fuselage). The fuselage prediction methods correlate well with test data over small (less than 5 deg) angles of attack but incur increasing errors with increasing incidence (errors to 30% or more at 20 deg angle of attack). No data are available relating to the estimation capability of the entire model. However, some estimation work at NADC has been successfully applied to the whole of the P.1127 Kestrel indicating that these methods have potential worth.

The only reasonable reliable method for estimating basic aerodynamics is the wind tunnel. Tunnel testing has indicated that unless extreme care is exercised with respect to model sizing, propulsion similitudes, and wall corrections, large discrepancies can be expected to result.

Although some of the techniques that were developed for estimating the basic aerodynamics of jet-lift aircraft may be applied to fan configurations, there is a fundamental difference between the two. Fans-in-wing incur a large separation of flow aft of the fan which makes theoretical prediction extremely difficult. The placement of fans combined with a change in wing planform can have a significant effect on aerodynamics. For instance, there are significant differences in pitching moment between wing-buried fans on a moderately swept wing and wing-mounted nacelle-buried fans on wings of large sweep. The point to be made is that geometrical configuration has a large effect on estimation capability. A two-dimensional theoretical method has been developed for calculating fan-in-wing lift. It is not known how well this technique compares with measurements.

Severe discrepancies can result in the measurement of lift for two differently scaled models of fan-in-wing aircraft. In the one case, the differences measured were so large at low inlet-to-jet velocity ratios as to render the estimations virtually useless since neither could be proved to be more accurate.

Estimation capability (both theoretical/semi-empirical and wind tunnel measured) can be expected to be far more reliable for tilt-prop/rotor and tilt-wing aircraft. Years of helicopter technology development account for this capability.

12. Technology Area - Static Stability

    Estimation Capability

    - theoretical/semi-empirical
        IGE: none
        OGE: poor-none

    - empirical-model
        IGE: none-poor
        OGE: poor-good

    - empirical-rig
        IGE: poor
        OGE: good

    Measurement Capability

    - flight test
        IGE: good-poor
        OGE: good

    Remarks

    - static stability depends on same parameters as trim
    - flight-test measurements involve the same shortcomings as for trim; fixed operating point testing is not always practical owing to limited sideslip envelop, lift engine operating time limits, etc.

Technology Summary

In addition to providing trimmability, static stability is related to airspeed, angle-of-attack and sideslip dynamics. For sideslip excursions, the pilot is also concerned about dihedral effect (variation of rolling moment with sideslip angle) and his ability to trim involves both directional and lateral control. Although a natural stable sense between controller deflections and velocity, angle-of-attack and sideslip angle is preferred, the condition is seldom if ever available in VSTOL aircraft, usually requiring considerable pilot attention to maintain trim while performing flight path maneuvers. Large instabilities markedly increase the pilot's workload and usually result in a poor opinion of the aircraft. The situation deteriorates considerably in the presence of gusts. Static stability is directly related to the accuracy of flight-path control throughout transition and is, therefore, of particular concern in tracking so-called "optimal" in-bound decelerating approaches.

Static stability is an integral function of the geometry of the aircraft and the effects attendant to the powered-lift system. As such, it must be estimated to reasonable levels of accuracy early in the design evolution. Since the aerodynamics are extremely difficult to estimate without some wind tunnel work, preliminary design must permit a wide margin of static stability, growth potential, or, conversely, a strict limit on stability degradation. Although some irregularities in static stability can be overcome through the use of augmentation systems and artificial feel/controller displacement characteristics, it is unwise to rely solely on such means for remedies late in the development cycle. Basic aircraft changes will be prohibitively costly at a late stage of development and will also incur a large delay in the program.

In hover and low-speed flight, the propulsion system (through direct and induced effects) dominates the character of static stability, regardless of VSTOL type. Static stability estimates require detailed knowledge of the power-induced flow fields in and about the aircraft, e.g., induced flows over tail surfaces. Such estimates have been of very poor quality, both IGE and OGE. Large variations occur over wide ranges of power level, angle-of-attack and sideslip. Augmenter concepts incur an additional burden when attempting to estimate pressure coefficient shifts with power, incidence and sideslip variations. In addition to propulsion and propulsion-induced effects, static forces and moments of consequence are generated independently through the action of steady winds and gusts acting upon the airframe. These static aerodynamics, although considerably easier to predict, cannot be treated meaningfully without inclusion of the attendant propulsive effects. Linear superposition of the two may suffice for "first-order" estimates, but generally will not be adequate for designing with confidence aerodynamic control surfaces and propulsive control devices.

Poor static stability can seriously reduce the effectiveness of IFR operations because of its effect on flight path tracking precision. Poor static stability will always place demands on more sophisticated VLA's, cockpit instruments and augmentation systems thus incurring an increased maintenance program. It adversely affects pilot training and will require more frequent pilot proficiency flights.

Static stability and trim depend principally on the same set of aerodynamic parameters. There is essentially no difference.

13. Technology Area - Control Power

    - trim
    - maneuvering
    - gusts
    - couplings

    Estimation Capability

    - theoretical/semi-empirical
        IGE: none-poor
        OGE: poor

    - empirical-model
        IGE: poor-none
        OGE: poor-good

    - empirical-rig
        IGE: poor
        OGE: good

    Measurement Capability

    - flight test
        IGE: none
        OGE: poor-good

    Remarks

    - estimation is less difficult for tilt-prop/rotor and tilt-wing VSTOL types
    - estimation often hampered by regions of unpredictable flow separation
    - power-induced local flows on tail and other control surfaces are unpredictable; therefore, control power is unpredictable
    - ground effect totally obscures theoretical estimation
    - actual vehicle hover control power is often computed from test stand data because response is nonlinear with control displacement and full control input is not practical, especially IGE

Technology Summary

Control power has been the single most discussed topic in technical papers, reports, and symposia on VSTOL stability and control. It deserves that position primarily because VSTOL aircraft control power is usually achieved at great expense to performance. In addition, the levels of control power finally designed into the aircraft often prove to be inadequate as revealed by the number of surprises in flight testing. Control power for a VSTOL aircraft must be adequate to trim the aircraft, maneuver, overcome gust upsets, and provide for controls coupling. Since these achievements are difficult to estimate accurately and since an abundance of control power is usually lacking, it is easy to see why the subject has received so much attention.

From a pilot's point of view, control power is inextricably associated with control sensitivity and aircraft damping. He "sees" control power through such things as being able to raise a wing in high sideslip conditions, trimming out an asymmetrical power situation, or providing an initial snappy response to commanded inputs. These capabilities and many more are directly sensed by the pilot as control power. They become primary factors in judging an aircraft's flying qualities.

Beyond the mid-development point, there is nothing that can be done to achieve a significant improvement in control power without incurring one or more of the following penalties:

- a large increase in cost
- a large delay in the program
- a compromise in performance

These penalties accrue regardless of the type of VSTOL aircraft, although they probably affect the higher disk loading aircraft more severely. Stability augmentation can do nothing to increase control power except possible to manage it better, thus conserving it for more important flight tasks. Sensing and responding to gust disturbances can be handled far more effectively in some cases through automatic means, thus making available more control power for maneuvering while at the same time freeing the pilot from his attention to the gusts.

Control power is very demanding on the propulsion system, increasing with increasing disk loading. The propulsive lift-generation system must provide adequate margins of energy to serve control power needs without compromising overall lift performance or vertical acceleration capability. Moreover, the propulsive lift system must also be responsive to the sporadic demands of the control system, a characteristic not easily achieved with some systems. Control power is also attended by a drag penalty when achieved either solely or partially by conventional aerodynamic control surfaces. This penalty is not too critical in hover and transition flights because of the low speeds involved. Good design of vanes, flaps, louvers, and the like must be a goal in achieving acceptable, aerodynamically efficient control devices/surfaces. Such surfaces often must perform under high downwash conditions and at high angles-of-attack. This area is associated with large flow separation phenomena and consequently poor estimation capability. The effectiveness of controls, regardless of form, can change rapidly in self-induced and ground-induced flow fields requiring a great deal of attention.

Control power is directly relatable to mission effectiveness, usually in terms of maneuverability. In spite of mission requirements, control power is a first order consideration in describing the flying characteristics of the aircraft. The ability of a VSTOL aircraft to station over a spot on the deck for on-loading, off-loading, or achieving a good position for a trapping device is a strong function of control power available and directly affects the efficiency of flight-deck management. Poor control power can lead to imposing efficiency-draining flight restrictions and certainly reduces pilot confidence in IFR operations.

Estimation of control power is found to be far less difficult for tilt prop/rotor or tilt-wing VSTOL aircraft than for others. Much rotor, propeller, and high-lift-devices technology is available to draw from. This does not imply, however, that estimates for such aircraft are necessarily accurate or even consistent, but rather that there is a broader technology base from which to derive useful design guidance.

14. Technology Area - Gust Sensitivity

Estimation Capability

- theoretical/semi-empirical
    IGE: none
    OGE: none-poor

- empirical-model
    IGE: poor
    OGE: good

- empirical-rig
    IGE: none
    OGE: none

Measurement Capability

- flight test
    IGE: poor
    OGE: poor

Remarks

- no known reliable theoretical or semi-empirical methods are available
- for flight test investigations, it is difficult to define gust profiles, especially in transition, and to quantify low aircraft speeds

Technology Summary

Gust sensitivity can best be described by viewing the moment increments imposed on the aircraft resulting from perturbations in speed largely occurring in the horizontal plane. Such perturbations (gust components of velocity) can have a marked effect on both the pitching moment (M) and rolling moment (L) of the aircraft. The derivatives,

- M_u - change in pitching moment due to change in forward perturbation velocity (u)

- L_v - change in rolling moment due to change in side perturbation velocity (v)

represent the most important contributions to this effect. Whereas for conventional airplanes with negligible power effects M_u is essentially 0, for VSTOL aircraft and their inherently large power effects M_u can be, and often is, a significant quantity. M_u has a great effect on VSTOL stability and increases in importance as disk loading decreases. An unstable M_u (negative, nose-down pitching moment with increasing speed) requires full attention on the part of the pilot in his attempts to hold a station in hover; the same is true for accelerated transitions. For tilt-wing aircraft, M_u is positive but decreases rapidly with increasing forward speeds. For ducted propeller aircraft, M_u is usually very large positively. Although a positive M_u is statically stable, a large value will quickly drive oscillatory dynamics into an unstable region. It is this form of instability combined with the attendant high-frequency components of horizontal gusts that create a sometimes untenable pilot control situation. The lateral case is analogous to that in pitch with one essential difference: acceleration command reduces even further the pilots ability to cope with side gusts.

Although it is possible to alleviate the effects of adverse gust sensitivity through automatic sensing and control, every effort should be made to reduce such effects initially through sound aerodynamic design. Some VSTOL concepts are inherently prone to gust sensitivity; little can be done except to assign a greater control power level to the augmentation system to overcome the effect and unburden the pilot. Too much control authority could, however, cause SAS (Stability Augmentation System) saturation, leaving the pilot with no maneuvering control margin. Regardless of how adjustments are made, adverse gust sensitivity if allowed to persist far into the development program will cause a deterioration in performance, increase the complexity of the control system, increase the cost of the program, and eventually the cost of the production aircraft as well.

By far, the technology most affected by gust sensitivity is that dealing with the aerodynamics of the propulsive-lift generation system; that is, ducted fans, tilt props, tilt wings, and fan-in-wings. Careful analysis of local flow conditions and predictions of momentum drag components and center of pressure migrations are important in estimating and consequently designing around gust sensitivity effects. Tradeoffs are often difficult to make since designs that may favor lessened sensitivity invariably reduce lifting performance.

Aircraft that are gust sensitive cannot be expected to hover with any operationally acceptable percision. In the same light, inbound flight path control will deteriorate thus reducing the efficiency of aircraft recovery. IFR capability will markedly diminish if the piliot must pay too much attention to gust correction in an already extremely taxing situation. Increased maintenance hours are also likely to keep more complex flight control equipment functional.

Very little can be said concerning estimation capabilities with regard to gust sensitivity terms. It is strongly suspected that of all the stability derivatives, those relating to perturbation velocities are most difficult to estimate theoretically or semi-empirically. Tunnel testing should yield reasonable results, depending upon the care exercised in sizing the model and providing good power similitude. Extraction of these derivatives from full-scale rigs is impossible because of the usual non-aerodynamic bird-cage configuration normally associated with such rigs and also because of inaccurate airspeed control and measurement. Deriving gust sensitivity parameters from flight tests would be equally difficult.

15. Technology Area - Coupling
    - Aerodynamic Controls
    - Controls Blending
    - Engine Gyroscopic
    - Aircraft Inertia

    Aerodynamic Controls

    Estimation Capability

    - theoretical/semi-empirical
        IGE: none-poor
        OGE: poor-none

    - empirical-model
        IGE: poor-none
        OGE: poor-good

    - empirical-rig
        IGE: not applicable
        OGE: not applicable

    Measurement Capability

    - flight test
        IGE: good
        OGE: good

    Controls Blending

Estimation Capability

- theoretical/semi-empirical: <u>not applicable</u>
- empirical
    - model : <u>not applicable</u>
    - rig   : <u>good</u>

Measurement Capability

- flight test: <u>good</u>

Engine Gyroscopic

Estimation Capability

- theoretical/semi-empirical: <u>good</u>
- empirical
    - model : <u>poor</u>
    - rig   : <u>good</u>

Measurement Capability

- flight test: <u>good</u>

Aircraft Inertia

Estimation Capability

- theoretical/semi-empirical: <u>good</u>
- empirical
    - model : <u>not applicable</u>
    - rig   : <u>not applicable</u>

Measurement Capability

- flight test: <u>good</u>

Remarks

- gyroscopic and inertial coupling are aerodynamically independent; estimation accuracy depends on accuracy of estimating maximum angular rates and moments of inertia only
- in-flight measurement of airspeed, angle-of-attack, and sideslip are inaccurate at low speed
- cross controls effectiveness is estimated with the same degree of difficulty as single-axis controls effectiveness

Technology Summary

VSTOL aircraft are notorious with regard to the number and types of coupling they display, such as controls cross-coupling, gyroscopic coupling, inertial coupling, and controls blending. These forms of coupling are readily apparent to the pilot. Controls cross-coupling is most noticeably apparent since off-axis trim is required to compensate for it. Such coupling is always annoying to the pilot, detracts from his main flight task, increases his workload, and results in a downgraded opinion of the flying qualities of the aircraft. As an example, modulation of thrust to achieve height control is often accompanied by moments and forces about and along other axes. Tilt-prop/rotor, tilt-wing, augmenter, and fan-in-wing VSTOL aircraft are more prone to this type of coupling and often simultaneously incur lateral as well as directional trim changes.

Gyroscopic coupling results from the rotating mass of the compressor and turbine for a jet-lift, the rotating mass of fan blades for a fan-in-wing, and other rotating masses for the remaining concepts. Use of contra-rotating masses can greatly reduce the effect, but such solutions are not always possible in a given design. Such coupling is particularly noticeable at low-speed high-power-setting conditions, especially when through either the pilot's command or an external disturbance the aircraft is rotated abruptly at a high initial rate. Gyroscopic coupling effects can require considerable pilot attention and usually result in an unacceptable pilot rating.

Inertial coupling is of far less concern for most VSTOL aircraft except for the jet-lift type, which inherently compacts an unusual amount of mass in the fuselage thereby yielding high moments of inertia about the longitudinal axis. Inertial coupling effects are prominent at high aircraft angles-of-attack when attended by rapid angular rates of motion. Again, the consequence is reduced pilot acceptance.

Controls blending is included here solely because it is the means by which to counteract all other forms of natural coupling. It is designed into the flight control system to overcome the adverse effects of inherent coupling (in addition to which it also "blends" aerodynamic and reactive controls during transition). The estimation capability for controls blending is dependent upon the estimation capability of the other coupling terms.

The designer must be aware of all forms of coupling early in the development program, since late corrections can seriously compromise control power necessary for major flight tasks. Gyroscopic coupling, associated as it is with the propulsive lift system, is virtually impossible to reduce once a system commitment is made. Changes to diminish the natural causes of all forms of coupling must be made early in the program to avoid incurring a prohibitive loss in time and money. Remedial changes also would most certainly be costly in terms of lift performance, control power, and flight control system complexity.

Gyroscopic and inertial coupling, being aerodynamically independent, can be estimated to a high degree of accuracy; such accuracy relies wholly on the quality of moments of inertia and maximum angular rate estimates. With regard to estimation capability, aerodynamic/propulsive coupling does not differ much from other control effectiveness terms. There are no reliable theoretical and semi-empirical methods for all VSTOL aircraft. Tunnel estimates, however, can be expected to yield a reasonable degree of accuracy if accepted procedures are followed.

16. Technology Area - Dynamic Stability
    - pitch
    - roll
    - yaw

    Estimation Capability

    - theoretical/semi-empirical
        IGE: none-poor
        OGE: poor

    - empirical-model
        IGE: poor
        OGE: good

    - empirical-rig
        IGE: not applicable
        OGE: not applicable

    Measurement Capability

    - flight test
        IGE: poor
        OGE: good

    Remarks

    - random nature of ground effects disturbances precludes extraction of modal response IGE

Technology Summary

The dynamic response characteristics of VSTOL aircraft are of greatest importance in tight flight tracking tasks. Unlike the classical linearized behavior often assumed for conventional aircraft, VSTOL aircraft do not necessarily respond in a manner linearly related to the control input. Nonlinear aerodynamic and propulsive effects, "designed-in" nonlinear control interconnects, and discontinuities in conversion from aerodynamic to reactive controls produce responses in the actual aircraft that can be far different from those predicted on the basis of linearized, possibly even uncoupled, equations of motion.

Pilot acceptance of VSTOL dynamics is a function of the type of control system at his disposal:
- acceleration
- rate
- attitude

The acceleration system is the simplest, but usually incurs the greatest pilot workload since he must provide his own stability and angular rate damping; that is, there are no stabilizing feedbacks. The rate system, on the other hand, provides rate feedback permitting attitude control within the altitude stability loop and freeing the pilot from worry about excessive rate buildup. The pure attitude system incorporates altitude feedback in addition to rate feedback. The pilot commands steady-state altitude proportional to stick deflection, and all stabilizing requirements are automativally provided. There are, of course, various combination of these systems that have been designed into some VSTOL aircraft.

The pilot is keenly aware of the character of the aircraft's dynamic response, adapting his own control element to supply the needed anticipation, rate estimation, smoothing, and sometimes a completely different control law in an attempt to achieve acceptable response. When dynamics are beyond his natural or willing adaptation, the pilot will downgrade the flying qualities of the aircraft.

Poor basic dynamics often persist in a VSTOL aircraft far into its development program. Simultaneously, a stability augmentation system is instigated to remedy the now hopeless basic airframe stability characteristics using piloted simulators and aerodynamic data. This practice continues because of a fundamental inability to predict VSTOL aerodynamics well enough to design better levels of dynamics into the aircraft at an earlier stage. Irresponsible and total disregard for dynamic behavior even at the preliminary design stage can lead to serious unsolvable problems, regardless of the sophistication of the augmentation system; in addition, such complexity is not desirable from a reliability standpoint. It is clear that when less is known about the natural dynamics of the aircraft, the program will be more costly to develop, including the rigs, simulators, and other associated resources needed to apply remedial correction.

Poor basic airframe dynamics cause a complex auto-stabilization system that may increase maintenance time and cost as well as aircraft downtime. In turn, this combination reduces operational effectiveness. Poor dynamics render the aircraft relatively ineffective as a stable platform for delivery of weapons, placement of ASW equipment, and on-loading/off-loading ship's cargo. Unstable aircraft will drive both cockpit displays as well as shipboard

VLA's into more sophisticated designs, causing additional cost, complexity, maintenance, and training of personnel. Pilot training will be more difficult. More frequent, costly, and time consuming hours will have to be allotted to maintain pilot proficiency.

Damping and rotary derivatives are among the most difficult to calculate as well as to determine empirically. For jet-lift and some fan-in-wing/body aircraft, a method for the estimation of component contributions to these derivatives has been developed. However, the component calculations must be carefully combined with highly tailored interference functions to assure reasonable confidence in the total aircraft values.

17. Technology Area - Height Control
    - Power
    - Damping

    Power

    Estimation Capability

    - theoretical/semi-empirical
        IGE: poor
        OGE: good

    - empirical-model
        IGE: not applicable
        OGE: not applicable

    - empirical-rig
        IGE: not applicable
        OGE: not applicable

    Measurement Capability

    - flight test: none

    Damping

    Estimation Capability

    - theoretical/semi-empirical
        IGE: poor
        OGE: good

    - empirical-model
        IGE: good
        OGE: good

    - empirical-rig
        IGE: not applicable
        OGE: not applicable

    Measurement Capability

    - flight test: good

Technology Summary

Height control has been separated from the normal disucssion of control power and dynamics because it represents a unique degree of control freedom peculiar to VSTOL aircraft and important enough to warrant special attention.

As with attitude control power, height control power must be sufficient to provide for

- trimming the aircraft under the most adverse loadings and wind directions
- gust upsets
- deliberate maneuvering
- cross-control to offset height coupling phenomena

Height control (power, damping and sensitivity) is a primary flying quality factor throughout transition and hover. The pilot "sees" vertical control and maneuvering quite differently for the various VSTOL aircraft types. With regard to tilt-prop/rotor, ducted fan, fan-in-wing/body, and tilt-wing (prop) aircraft, the pilot usually provides a collective command in a hover mode that, in itself, is a vertical-rate-response type of control with rate damping dictating the steady-state climb or descent rate. Jet-lift aircraft, in reality, provide a similar natural response mode of vertical control except that they usually exhibit far less vertical damping and the "character" of height control can be very different; engine time lags contribute much to this character. During transition, both collective (direct modulation) and vectoring of thrust takes place to provide vertical control; thus, second (and higher) order responses shape the character of height stability and control. In addition, the combination of direct modulation and vectoring is invariably accompanied by coupling, predominately about the pitch axis and along the longitudinal axis.

Since altitude and sink/climb rate control are critical to all hover and transition flight tasks and because they, in turn, are primarily dependent upon the character of height control, the pilot will regard such control as important in determining flying qualities acceptability.

Height control power is the most demanding of all axis controls with regard to the propulsion system. Since, for the most part, the propulsion system sets the pace of VSTOL aircraft development, height control and that portion of the total propulsive system energy output allotted to it becomes an important factor. Virtually no engine design flexibility exists to provide for significant increases in height control power allotment beyond the point of engine selection. Deficiencies are essentially impossible to overcome beyond a very early stage of aircraft development.

With respect to aerodynamics, height control relates primarily to height damping (both IGE and OGE) and secondarily to cross-axial control brought about through associated downwash effects. Height damping is inversely proportional to disk loading and directly proportional to planform area. Jet-lift vehicles historically expose far less normal area; consequently, they are doubly penalized with regard to a deficiency in natural vertical damping. Little design flexibility exists aerodynamically to augment vertical damping for jet-lift aircraft. However, preliminary design efforts should always attempt to provide as much of a natural "barn-door" effect as possible.

Vertical operational capability can be seriously compromised should height control be insufficient, imposing either severe flight restrictions and/or demanding larger flight deck areas for the accommodation of aircraft incapable of pure to near-pure "V" capabilities in either landing or taking-off. A deficiency in height control could also impose an undesirable penalty in TOGW as well.

V. PROBLEM AREAS AND ISSUES
INTRODUCTION

The preceding technology assessment addressed technologies having a significant interactive effect on VSTOL aircraft size and performance. This section covers those areas having a less mutual dependency, e.g., materials technology, avionics, air-sea interface, etc. Although this section is still part of the technology assessment, predictive and measurement capabilities are not treated, nor are they appropriate. Technological features, within the framework of component areas, are addressed here. Problem areas and issues are highlighted as they relate to major deficiencies impeding the progress of VSTOL aircraft development. The areas covered are:

<u>Concepts, Configuration and Systems Integration</u>
- Systems Considerations
- Aerodynamic Issues
- Control, Stability and Flight Dynamics Considerations
- Ground Effects
- Propulsion Induced Effects
- Aircraft-Ship Interface
- Reliability and Maintainability

<u>Propulsion Systems</u>
- Propulsion Technology Comparison
- Reliability
- Critical Propulsion Technologies

<u>Vehicle Stability and Control</u>
- Control Characteristics
- Vehicle Aerodynamics
- External Environment
- Stability Augmentation System
- Other Important Issues of the Control System

Materials and Structures
- Characteristics of Composite Materials
- Failure Modes
- Candidate Materials
- Manufacturing Technology
- Design Criteria
- Composites in Propulsion Systems

Avionics
- Flight Control
- AEW Radar

The format in this section is also noticeably different. Because of the large number of areas to be covered and to hold the discussion within reasonable bounds, a greatly abbreviated presentation is used. A "briefing" style of presentation has been adopted. Narrative explanation is thus sacrificed for brevity; but, it is hoped that the main thrusts are not lost. Again, major issues and problems are highlighted within each area.

V.1-CONCEPTS, CONFIGURATION AND SYSTEM INTEGRATION
INTRODUCTION

30 Years of VSTOL

- over $1.25 billion expended
- large number (about 55) of concepts/configurations investigated
  - jet-lift
  - fan-in-wing/body
  - ducted prop
  - tilt prop/rotor
  - tilt wing
  - deflected slipstream
  - augmenter
- demonstrated varying degrees of feasibility
- abundant technical problems
- lack of mission requirements

Perkins Committee (1960)

"The state-of-the-art of vertical and short takeoff landing aircraft has advanced to the point where VSTOL aircraft capable of meeting operational requirements can be developed. The full military usefulness of VSTOL aircraft must now be demonstrated through operational evaluation. Unless a program of operational suitability is initiated, the state of uncertainty that exists today will continue."

H.A.S.C. Hearings (1964)
- technical feasibility of VSTOL...has been clearly demonstrated
- ...there are valid military requirements for VSTOL...
- ...none of the present DoD programs are adequate to determine the the operational suitability of VSTOL...

AGARD/NATO Meeting (1969)
- VSTOL Aircraft and Their Applications
- Reaffirmed findings of Perkins Committee

VSTOL-CTOL Comparisons
- clear and easily quantified disadvantage for VSTOL when compared one-on-one with CTOL
  - -weight
  - -range
  - -payload
- VSTOL advantages difficult to quantify (more subtle)
  - -systems analyses and operational experience required
  - -reduce vulnerability through dispersion
  - -faster reaction time
  - -cost effective solution to advancing Soviet naval threat

Results
- large deck carrier: has served as CTOL reference for almost 50 years
- preference for CTOL: little enthusiasm for VSTOL in an era of super-carriers and Mach 2 fighters
- reluctance to establish VSTOL missions
- none of the early VSTOL concepts went much beyond experimental flight testing

VSTOL Today
- missions being established: air-capable Navy
  - -AEW (Airborne Early Warning)
  - -ASW (Anti-Submarine Warfare)
  - -SAR (Search And Rescue)
  - -MA (Marine Assault)
  - -COD (Carrier On-board Delivery)

- missions are driving vehicle concepts: previously vehicle concepts were searching for a mission
- commitment of resources
- dedication/enthusiasm exits (although certainly not universal)
- technological and operational optimism
- many severe problems still exist

V.1.1-SYSTEMS CONSIDERATIONS

Systems Tradeoffs

- VSTOL range, speed, weight, payload, loiter time
- number of dispersed air-capable ships/size
  -smaller ships: more dispersion and less sea area to cover
  -shorter range, smaller aircraft
- small ships
  -less deck and hangar space
  -fewer aircraft
  -less maintenance and logistics support available
  -higher aircraft reliability requirements
  -more severe ship motion in heavy sea

Performance

- dependent on total system characteristics
  -ships
  -aircraft
  -logistics
  -command and control
  -
  -
- total system not adequately defined: VSTOL still does not have a defined reference

Operations Research/Systems Analysis

- must be used throughout design process
- optimize total system
- identify hierarchy of parameters driving system effectiveness
- influence of aircraft/ship constraints
- analyze mission requirements: optimum aircraft/ship combination for maximum mission effectiveness

V.1.2-AERODYNAMIC ISSUES

Engine/Airframe Integration

- high bypass lift/cruise engines placed in large diameter nacelles close to fuselage for minimum folded size

- satisfactory for low speed operation
- less than satisfactory for cruise: large protuberances in airstream
- decrease the lift-to-drag ratio
- configuration should be arranged to provide good low speed/high speed compatibility from lift and drag considerations
- must be resolved in wind tunnel tests

Transition Corridor

- ability to make steep descents with reduced power: a plague to most VSTOL concepts
- duct lip stall on configurations where engine pods are tilted to achieve conversion: reduces performance and induces high blade stresses
- jet induced lift and pitching moments as they affect transition
- resolve in wind tunnel tests

Dihedral Effect

- rolling moment in sideslip at low forward speeds
- severity increases with angle-of-attack
- loss of several VSTOL aircraft
- must be resolved by wind tunnel testing

Ground Effect Problems

- aerodynamic lift losses
- induced moments
- roll and pitch trim requirements
- hot gas ingestion
- must resolve in wind tunnel tests

Thrust Vectoring

- improve the efficiency of the turning process

Gear Boxes, Shafting, Clutches

- minimize for reliability purposes

V.1.3-CONTROL, STABILITY AND FLIGHT DYNAMICS CONSIDERATIONS

Air-Sea Interface

- formidable problem: control of VSTOL aircraft under all weather conditions (sea state 5) while landing on a rolling, pitching and heaving deck with limited visibility

- wind gusts
- stack gas flow fields
- induced wakes from ship superstructure
- suck down effects
- instability of "fountain effects" during aircraft/deck motions

Quote

". . .a challenge, but one which can be met using principles and technology that are emerging."

Claims

- VSTOL potentially easier to land under IFR (Instrument Flight Rules) conditions and on a moving deck than CTOL aircraft
- flight control is easier because vertical velocity is independent of forward velocity
- with a vertical flight capability, landing can be delayed as required once the aircraft is positioned over the deck and landing effected at pilot's or safety officer's discretion
- a VSTOL can move backwards and sidewards as well as forward to more easily correct position errors

Needs

- very sophisticated control system with high degree of automation
- high levels of control power
- rapid response thrust modulation to produce control moments about all three aircraft axes
- drift correction along each axis
- inertial reference system with three axes stabilization: aircraft referenced to landing spot and feedback system keyed to ship motion
  -land when deck is horizontal
  -land at any point in ship's roll
- human factors: HUD (Head-Up Display) required showing deck position relative to aircraft and height to go

Implications

- digital flight control and engine control systems incorporating digital fly-by-wire control actuators
- high reliability

- triplex redundant on-line failure
- aircraft T/W ratios greater than 1.05 with sufficient reserve power (thrust margin) to produce control accelerations about all three axes and provide adequate vertical control in takeoff and landing

Issues/Problem Areas
- amount of trim control required
- amount of required thrust margin
- control power and sensitivity requirements for all axes including height control
- control mechanization
- degree of sophistication needed for three-axis stability and control augmentation systems
- issues must be resolved through piloted motion simulators and adequately equipped flight research aircraft

V.1.4-GROUND EFFECTS

Suckdown Effect: downwash from VSTOL aircraft spreads from the impingement point and the outward flowing sheet produces suckdown forces (negative lift)

Fountain Effect: flows from multiple lifting units meet and produce an upward flow which impinges on the bottom of the aircraft resulting in a lifting force (positive lift)

Net Lift Increment: dependent on configuration and can be either positive or negative
- aircraft should be configured to maximize favorable fountain flow effect
- favorable fountain flow, however, can be modified by aircraft/ship deck attitude and can produce overturning moments on aircraft

V.1.5-PROPULSION INDUCED EFFECTS

Aerodynamic Lift Loss and Asymmetric Lift
- caused by exhaust/inlet flow upon aircraft surfaces
- dependent on aircraft configuration
- analytical and experimental test techniques need improvement
- prediction techniques are critical

Propulsion System Effect on Flight Dynamics

- very serious: adverse influence on roll and pitch trim requirements
- assessment of cross flow effect on control trim cannot be made
- prediction techniques need to be improved
- large scale wind tunnel tests need to be conducted to accurately establish control moment requirements for all aircraft axes

Recirculation and Hot Gas Ingestion

- prediction techniques need improvement
- difficult to establish judicious selection of engine inlet location and exhaust gas direction
- can seriously reduce liftoff and landing performance

V.1.6-AIRCRAFT/SHIP INTERFACE

Harrier (AV-8A) Operation

- cleared to operate from carriers and LPH-2 and LPD class ships
- one mile visibility
- 400 ft. ceiling
- sea-state 3 (3 ft. to 5 ft. waves)
- except for daytime VFR conditions, the task is a high workload situation, not desirable for continual operations by fleet pilots

XC-142A and CL-84 Operation

- both operated aboard ship with relative ease . . .
- under daytime, calm sea conditions

Operation Summary

- experience with fixed wing VSTOL shipboard operations is very limited
- still refining the helicopter/small ship "dynamic interface" operational procedures

Key Factors

- control power requirements, including minimum thrust-to-weight ratio:

  ```
  effected by
  -ship's turbulence (stern and island)
  -ship motion
  -wind-over-the-deck
  -pilot cues available: visual landing aids, head-up display, deck
   motion indicator
  -stabilization: both aircraft and ship
  ```

- HUD (head-up display)
  - essential for high-workload environment: weather/night ship landings
  - clear, logical and uncluttered
- close-in guidance system
  - automated: degree of automation is unresolved
  - relieve pilot workload
  - role of landing signal officer has to be determined
- visual landing aids
  - flexibility of operations
  - avoid requirement for ship to maintain specified heading during recovery operations
  - helicopter visual landing aids may not be applicable
  - night lighting systems must be developed for each class of ship
  - pilot depth-perception not reliable at night
- deck securing system
  - requirement for and system not defined
  - aircraft configuration can impact requirements
- winching-in systems
  - requirement for and system not defined
  - may impose unusual loads on landing gear: high side loads when contacting a pitching, heaving deck
- deck traversing system
  - requirement for and system not defined
  - moving VSTOL aircraft into tight fitting hangar on the deck of small ships in high sea states is a formidable problem
- fin stabilization of small ships
  - reduce rolling motions in high sea states
  - program needed to investigate tradeoffs of ship fin stabilization and aircraft handling gear

Summary

- ship/aircraft interface environment is severe and the problems as outlined above must be properly considered
- total system approach, including the ship, is required

V.1.7-RELIABILITY AND MAINTAINABILITY (R&M)

R&M Presently Achieved (newest Navy tactical fighter)

- MFHBF*: 1.0 - 1.5 hr.
- MTTR**: 1.6 - 1.8 hr.

R&M VSTOL Goals

- MFHBF: 11.0 hr. (min.)
- MTTR: 1.0 hr. (max.)

---

* Mean Flight Hours Between Failures

** Mean Time To Repair

Issues/Problem Areas

- R&M is extremely critical because of the dispersion of small numbers of aircraft per air-capable ship and the maintenance facilities, logistics support and service crew are limited on smaller ships
- high R&M goals drive up aircraft weight and cost
- tradeoff studies are required on the effect of R&M on aircraft weight, size, cost and logistics support
- are the above R&M goals realistic?: they are many times higher than current practice requires or supports

Keys

- simplicity, modularity, interchangeability
- use simplest propulsion system possible
- limit turbine inlet temperatures
- reduce number and complexity of shafting, clutches and gear boxes

V.2-PROPULSION SYSTEMS

INTRODUCTION

Current Engines

- will not meet minimum performance requirements for VSTOL

Needs

- higher thrust-to-weight ratio than fighter aircraft
- better specific fuel consumption than transport aircraft
- higher reliability than normal
- readily accessible
- easily maintainable
- high diagnostic capability
- higher output engines per unit weight and per unit volume
- increased use of composite materials and structures

Critical Issues/Problem Areas

- inability to predict engine/airframe integration effects in assessing aircraft performance potential, particularly in vertical and transition modes
- propulsion system is vulnerable to shortfalls or losses in reliability

and must absorb all shortfalls, including its own, as well as any added load imposed by additional equipment
- failure to absorb these shortfalls compromises the aircraft
- a "thrust pad", i.e., a thrust capability above the specified value, must be provided which can be traded for shortfalls

V.2.1-PROPULSION TECHNOLOGY COMPARISON

	Today	VSTOL Goal
Bare Engine Thrust/Weight (T/W)	5-7	9-11
Turbine Inlet Temperature (TIT)	2400°F	2700°-2900°F
Specific Fuel Consumption (SFC)	.6	.6 with high thrust ratio
Core Compressor Ratio	10-14	14-22 on single spool
Fan	fixed geometry	variable geometry
Reliability MFHBF*	63 hrs.	same for more complex system

*Mean Flight Hours Between Failures

- the 300°F to 500°F increase in TIT gives about 22% to 36% higher static thrust: the balance of the increase in T/W for bare engine must be produced by

    -higher loadings for stages of compressor and turbine
    -reduce number of stages and blades
    -extensive use of composite materials
    -reduce weight of gears, clutches, bearings, shafts in lift fans

- composites are extensively used in NASA QCSEE (Quiet, Clean, Short Haul, Experimental Engine)

    -fan blades
    -compressor fan frame
    -nacelle
    -inner and outer duct for by-pass air
    -nozzle flaps

V.2.2-RELIABILITY

Propulsion Systems

- 2, 3 or 4 core engines

    -connected to fans directly
    -connected through main reduction gears
    -connected through cross-shafting
    -connected by gas distribution ducts for tip driven turbine fans

- 2, 3 or 4 fans

    -fixed inlet guide vanes (FIGV) with fixed rotor blades
    -variable inlet guide vanes (VIGV) with fixed rotor blades
    -variable pitch (VP) rotor blades

Design Alternatives

- increased emphasis on durability/reliability, in general
- VSTOL goals are formidable
- design to reliability
- design to performance
- design to cost

High By-Pass Ratio Engines

- experience in wide-body commercial aircraft illustrates the difficulty in obtaining reliability while seeking high performance
- large increase in maintenance

Issues/Problem Areas

- cost implications of design to reliability

    -large amount of hardware
    -extensive operating time
    -double the "usual" development cost

- considering the complexity of VSTOL propulsion systems, are the reliability goals realistic?
- logistics of a VSTOL aircraft requiring a new engine while on a small ship

V.2.3-CRITICAL PROPULSION TECHNOLOGIES

Cycles - Airframe/Propulsion Integration
- optimum power plants for various arrangements
- tradeoffs between
  - oversized engines for emergency engine-out versus contingency ratings
  - increased number of engines versus contingency ratings
- influence of design and off-design performance with signficant amounts of bleed
- influence of unsteady flows when hovering IGE and during transition: engines with high tolerance to flow distortion (both velocity and temperature)
- capability of engines to operate in an abrasive environment: ingestion of dirt, dust, sand, other foreign objects

Fans
- variable inlet guide vanes (VIGV): advantages and disadvantages of thrust control by fan flow modulation cannot be made
- variable pitch (VP) fans: data base is larger but needs augmentation
- advantages of VIGV fans:
  - stability
  - reliability
- advantages of VP fans:
  - efficiency
  - flow control
- weight: a standoff between VIGV and VP fans
- issue: risk involved in VP fan when permissible tip speed may be lowered because of the pitch change mechanism (and weight) which are functions of tip speed

Power Transmission
- upwards of 17000 hp required during takeoff and landing
- propulsion systems using gears, clutches and shafting must transmit this power
- reliability: the simplest are the most attrative since many gears, clutches and cross-shafting are reduced, as well as weight
- issues

- components have not achieved the required levels of high power transmission at minimum weight
- risks created in the shafting and gearing by aeroelasticity of mounting
- readiness factor implications on component weights

Digital Engine Control

- the engine plays an important role in stability, control and flight qualities of VSTOL aircraft
- there is considerable activity in engine control by digital equipment, but no operational engine uses digital control
- for VSTOL, engine control should be integrated into a combined system that will control both engine and the aircraft
- issue: if engine control is made a function of a combined system (central computer) where is responsibility for the life and operation of the engine placed?

Composite Engine Components

- to achieve the required thrust-to-weight ratio of base engine, the structural weight of the engine must be reduced
- QCSEE (Quiet, Clean, Short Haul, Experimental Engine) uses composites in the inlet, fan frame, fan blades, inner and outer duct material and variable area nozzle
- little work in use of composites at other engine locations
- issue: can composite fan blades be developed that can meet the requirements of large bird strikes?

Core Engine Compressor

- performance characteristics

  - single shaft
  - significant pressure ratio (14-22)
  - large amounts of varying bleed flows
  - variable stators and inlet guide vanes to provide wide range of flow variations
  - high stage loading
  - efficient and stable operation

- no compressors with above combined performance charactertics have ever been built
- issues: reliability and weight

V.3-VEHICLE STABILITY AND CONTROL

INTRODUCTION

Formidable Problem

- VSTOL operational requirement to takeoff and land on a small moving deck, subject to diverse wind disturbances, in bad weather (sea state 5), at night
- this operation presents an extremely complex dynamic control problem
- hover and transition are most demanding parts of flight envelope
- shipboard operational experience limited to
  - Harrier
  - XC-142A
  - CL-84
- VFR (Visual Flight Rules) experience
  - only under daytime conditions
  - in relatively calm sea
- flight path control: greater precision is required when operating from a moving platform
- pilot workload: very large because of unsatisfactory handling qualities and inadequate display information

Essential Requirement

- land in a completely automatic mode
- provide sophisticated stability augmentation system (SAS), backed by appropriately designed displays

Issues/Problem Areas

- control system, displays and guidance interfacing
- interaction and cooperation between airframe/engine and control system designers
- obtain sufficient aerodynamic data on configurations to determine control power requirements for trim and upset
  - too much control power: weight and performance of aircraft are penalized
  - too little control power: aircraft unsafe and unable to meet mission requirements
- digital flight control system necessary because of high reliability requirements
  - fly-by-wire
  - triplex redundant
  - on-line electronic failure management

- detailed dynamic understanding of forces and moments created by aerodynamic and induced forces is not available
- design of high performance control systems is strongly influenced by prior knowledge of accurate dynamic models of the engine itself
- redundant engine sensors: control system must be tolerant to sensor failures since their failure rate may be unacceptable high during lift-off and landing when maximum thrust is required

V.3.1-CONTROL CHARACTERISTICS

Control Power

- required for

    -control of aircraft about all three axes
    -translational control
    -height control

- plus an adequate margin for

    -trim
    -upset
    -maneuver

- trim requirements

    -large amount of control power needed to compensate for trim and upset IGE
    -trim requirements depend on configuration requirements
    -there is lack of predictive capability for low speed aerodynamics for various configurations
    -therefore, trim requirements cannot be predicted
    -hence, wind tunnel experiments on detailed models are required to determine trim requirements

- issues/problem areas

    -lack of predictive capability for trim requirements IGE
    -effect of uncoupling the attitude/translational modes: simulator and flight studies are required

Engine Dynamic Response

- as aircraft hovers over a moving deck, there are large aerodynamic forces due to proximity effects that must be compensated for in order to land safely
- control during hover and transition depends heavily on ability of propulsion system to rapidly generate forces and moments for attitude control
- generation of control moments is dependent on configuration
- if bleed air is not used for attitude control, the control moments must be generated by the engines through the variation of thrust

- if bleed air control is used, control moments will require a large percentage of bleed air from the engine, thus requiring the development of a variable cycle engine
- the advantages of variable inlet guide vanes or variable pitch fans for generating the necessary thrust modulation are not clear, especially given the uncertainties of the aerodynamic ground effects upon the engine itself
- issues/problem areas
    - lack of experimental data on dynamic response characteristics of advanced engines
    - limitations on engine response time due to turbine temperature limits
    - limitations on engine response time due to inertia and stiffness of drive systems
    - gas coupled systems require more development in the control area
    - detailed dynamic data are required on engine/fan response characteristics
    - impact of extensive transient engine operation at high power levels on engine life and reliability
    - trade-offs between reaction controls and engine controls for moment generation

Independent Force and Moment Controls

- fundamental questions relating to landing a VSTOL on a rolling pitching ship:
    - whether to provide independent translational force controls in addition to moment controls so that the aircraft can be translated laterally without rolling?
    - if the rolling deck must be followed, is it necessary to decouple roll and side translation?
    - in the longitudinal direction, would the necessity for independnet axial force controls determine the tilt rate of the thrusters (depending upon configuration)?
- issue/problem area: answers to above questions must be resolved by simulator studies

Thrust-to-Weight (T/W) Ratio

- T/W margin for vertical control has an important impact on aircraft/engine design
- T/W = 1.05 appears marginal for vertical take-off and landing
    - uncertainties as to the actual value of the installed thrust of the engine
    - need to cope with a heaving deck
- the value of T/W = 1.05 is also marginal for one engine out emergency
- issue: should higher thrust margin be specified until operational experience defines an allowable lower limit?

V.3.2 - VEHICLE AERODYNAMICS

Stability Characteristics

- VSTOL aircraft have poor stability characteristics at low speeds due to
    - low levels of angular damping
    - little directional stability
    - high dihedral effect
    - little angle of attack stability
    - sensitiveness to gusts due to large nacelles/inlets
    - sensitiveness to exhaust/aircraft interactions
    - sensitiveness to ground effects: small details of the configuration can be of considerable significance

Needs

- experimental data on detailed configuration scale models
- data on trim characteristics in all possible symmetric and asymmetric flight conditions (emphasis on low speeds)
- transition corridor limits must be explored and sufficient control power provided to trim throughout the low speed envelope
- adequate margins for maneuvering must be provided to prevent upset from large external disturbances

Issues/Problem Areas

- analytical predictive capabilities and experimental data are required in all of the above areas for each configuration
- direction of the mean wind can change by as much as 60 degrees in about 20 feet of vertical displacement: the effects of such natural wind changes upon the aircraft are not understood and are strongly related to aircraft configuration

V.3.3 - EXTERNAL ENVIRONMENT

Sources of External Disturbances

- ship superstructure: produces large unsteady wakes
- stack gases
- relative wind

Ship Motion

- lower in frequency compared to bandwidth of aircraft control system
- magnitude and frequency of ship motion is random
- difficult to predict future motion from present state

Issues/Problem Areas

- all VSTOL configurations are sensitive to large flow disturbances in final approach
- above effects need to be quantified

    -inputs to simulation studies
    -required for aircraft control system design
    -modify ship superstructure shape to reduce wake disturbances
    -assess importance of ship stabilization with respect to aircraft control system design

V.3.4-STABILITY AUGMENTATION SYSTEM (SAS)

Needs

- some level of stability augmentation, as well as a display, must be provided for the pilot during landing
- simulation studies indicate that IFR landings on a small ship in sea-state 5 could be accomplished with acceptable pilot workload provided that

    -a translational velocity command control system is available
    -command display information is available

- there is a strong need to validate simulator results in flight

    -to optimize the control system characteristics
    -to avoid severe compromises in performance

Issues/Problem Areas

- how to best provide path guidance information?
- what control laws to use through transition to landing?
- degree of automation of landing task with respect to cost, complexity and reliability?
- desirability of having the pilot participate?
- research carrier aircraft is required to provide a realistic evaluation of the control system and pilot displays developed with the simulator studies

V.3.5-OTHER IMPORTANT ISSUES OF THE CONTROL SYSTEM

Air Data Requirements

- to what extent are air data sensors required as inputs to the flight control system?
- is it necessary to measure sideslip angle or is lateral acceleration satisfactory?

- due to complex flow fields that exist at low speeds, it may be difficult to locate air data sensors on many configurations of VSTOL aircraft
- control system/aircraft dynamics studies are required to resolve this issue

Guidance-Ship Motion Coupling
- how should the motion of the aircraft be coupled to the ship?
- how would the relative motion be sensed?
- how would the ship's motion be presented to the pilot, i.e., present state of the ship, predicted state, etc?

Flight Control Computer
- centralized so that all inputs from engine and vehicle motion are provided at one location and all actuator outputs are sent from one location
- provide for future growth/modification

Control System Design Interaction
- development of a satisfactory control system requires a closely coordinated relationship between the
    - configuration/aerodynamic designer
    - propulsion designer
    - control system designer
    - avionics designer

V.4-MATERIALS AND STRUCTURES

INTRODUCTION

Background
- metals inherently cannot meet VSTOL goals
- for the first time, an aircraft is viable only through extensive use of advanced composite materials
- aircraft and engine structure must be optimized in terms of strength-to-weight in order to achieve maximum overall weight reduction
- comparison of usuage of advanced composites
    F-18: 10%
    AV-8B: 20%
    VSTOL: 50%
- an attainment of a 50% usage factor will be an appreciable advancement

Issues/Problem Areas

- composite materials maufacturing techniques now require a large amount of hand labor: automation is required to reduce cost and increase speed of cutting, fabricating and assembly
- the diverse and unfamiliar failure modes of composite materials should be investigated and theories validated
- a program is needed for application of advanced composites in propulsion systems

V.4.1-CHARACTERISTICS OF COMPOSITE MATERIALS

Definition

- two or more materials each performing its function to provide the desired strength and stiffness
    - graphite/epoxy
    - boron/epoxy
    - boron/aluminum
- above examples of fibrous composite materials use graphite and boron fibers imbedded in an epoxy resin or an aluminum matrix
    - fibers: contribute strength and stiffness
    - matrix: provides the medium to support the fibers and hold them together

Advantages of Composites

- high strength-to-weight ratio
- high stiffness-to-weight ratio
- can be tailored to provide desired strength and stiffness required by application
- improved fatigue resistance
- resistant to environmental effects
- vibration damping
- low coefficient of friction
- low thermal coefficient of expansion

Comparison of Graphite/Epoxy Composites With Metals: see Table V-4-1.

Observations

- uni-directional graphite/epoxy
    - four times stronger per unit weight than the best aluminum, titanium and steel alloys
    - three times better in stiffness per unit weight

TABLE V-4-1. COMPARISON OF GRAPHITE/EPOXY COMPOSITES WITH METALS

	Density(lb/in^3)	Ultimate Tensile Strength(lb/in^2) 10^3	Young's Modulus(lb/in^2) 10^6	Strength-to-Weight(in) 10^6	Stiffness-to-Weight(in) 10^6
Graphite/Epoxy[1] Uni-Directional	.055	210	18.5	3.800	336
Graphite/Epoxy[1] Isotropic Lay-up[2]	.055	82	7.2	1.500	131
Alum. Extrusion 7075-T6[3]	.101	85	10.3	.842	102
Titanium Plate 6Al-4V[4]	.16	130	16	.813	100
Steel Forging D6ac[5]	.283	220	29	.777	102

1. Type 3501/AS
2. [0/± 45/90]
3. used in C-5A wing
4. used in F-14 wing
5. used in F-111 wing

- metal alloys do not differ significantly from each other on a per unit weight basis
- graphite/epoxy (isotropic lay-up): the most inefficient manner in which the filamentary composite can be used, but still
  - the material is almost twice as strong as metals
  - 30% stiffer than metals
- advanced composite materials have the inherent properties to bring structural weight fractions down to the level required for viable VSTOL aircraft

An Example - A Wing

- metals, being isotropic, have the same strengths and stiffnesses in all directions
- but loads on wing are such that the spanwise demands are much larger than chordwise demands
- hence, a metal wing is much stronger in the chordwise direction than it needs to be
- using composites in wing would reduce the weight and put the strength and stiffness where it is needed, i.e., in spanwise direction

V.4.2-FAILURE MODES

Background

- failure modes are diverse and unfamiliar due to anisotropic nature and inhomogeneity of composites
- failure characteristics of composites used in structures are currently assessed on a case-by-case basis by experimental means
- it is expensive to develop laminates by testing

Issues/Problem Areas

- there are no proven theories available which can design, a-priori, a laminate with the same level of confidence which exists in the design of metals
- formulation and verification of failure theories are required

V.4.3-CANDIDATE MATERIALS

Graphite/Epoxy

- most experience has been with these systems
- there are several brand names with sufficiently different properties to negate interchangeability

Boron/Epoxy

- airframe
- rotating components of turbo-fan engines

Kevlar/Epoxy

- will most likely replace glass/epoxy in secondary structures because of its much lower density
- its inferior compressive properties negates application to primary structures

Issues/Problem Areas

- composite materials technology is still maturing
- ease of fabrication needs to be improved
- moisture resistance needs improvement
- proliferation of material systems

V.4.4-MANUFACTURING TECHNOLOGY

Background

- fabrication of composite materials into a structural component is a "building-up" process wherein piles are layed-up to the desired thickness
- "build-up" process requires relatively light machinery whereas metal processes (stamping, forging, machining) require heavy and massive machinery

Issues/Problem Areas

- automation for fabrication of advanced composite materials is presently at a primitive level compared to metal technology
- the best or optimum methods of fabrication remain to be developed
- R&D dollars must be applied to overtake the 50 years of metal technology

V.4.5-DESIGN CRITERIA

Background
- current safety factor is 1.5 (to insure structural integrity)
- 1.5 was chosen because it is the ratio of ultimate stress to yield stress for an early aluminum alloy
- advanced composite materials do not possess a "yield stress"; the stress-strain relation is essentially linear to failure

Issues/Problem Areas
- it is not known whether the 1.5 factor of safety is too small or too large for composites
- since structural weight is directly related to the factor of safety, this requires further study

V.4.6-COMPOSITES IN PROPULSION SYSTEMS

Background
- in view of the obvious need to reduce the empty weight of the aircraft, the development of advanced composites for propulsion systems should be pursued

Potential Areas for Application
- shafting which connects fans to engines and cross-shafts between fans
    - translation into hardware needs to be implemented
    - need to gain service experience in a realistic environment
- the large necelles which are wrapped around the fans and engines
    - size may be a problem
    - new structural concepts may be required
    - major uncertainty is in the nature and magnitude of the aerodynemic loads on nacelles
- fan blades
    - pacing problem is the FOD (foreign object damage) requirement of bird strikes
    - most promising approach to bird ingestion damage is a hybrid design using boron - aluminum composite as the aerodynamic shell which is bonded onto a titanium spar

Issue/Problem Areas
- a program is needed for application of advanced composites in propulsion systems

V.5-AVIONICS

INTRODUCTION

Background

- avionics encompasses all onboard electronic and electrical systems and includes electro-optical subsystems (avioptics)
- an avionics suite consists of a common subsystem for flight control and subsystems for mission-peculiar designs, e.g., AEW, ASW
- avionics/avioptics systems for AEW and ASW missions are "design driving" installations which determine, in large part, the aircraft
    - configuration
    - size
    - weight
    - cost
    - reliability and maintainability
    - operational availability
    - logistic support systems
    - IOC date

Flight Control Avionics (see also Vehicle Stability and Control)

- full authority digital electronic control system incorporating digital fly-by-wire control/actuators
- triply redundant on-line failure management
- advanced integrated cockpit display including a suitable heads-up display (HUD)
- digital data processor
    - commonality
    - interchangeability
    - central/modular/programmable
    - fiber optics busses
    - LED sources and connectors

Mission-Peculiar Avionics

- dominated by the AEW radar which effects the entire aircraft design
    - antenna weight, size, shape, aerodynamics
    - antenna type: rotodome, radome, phased arrays (planar or conformal), "fin" arrays
    - frequencies: UHF, S-Band or L-Band
    - RF power generation, packaging, cooling
- operational mode
    - SAR (Synthetic Aperature Radar)
    - OTH/DCT (Over-The-Horizon Detection, Classification and Targeting)
    - electronic agility

Issues/Problem Areas

- the current diverse avionics programs do not tie together to form one complete and integrated/coordinated program
- a realistic EW/ECM threat must be defined for the 1990-2010 period, against which the AEW mission can be designed
- tradeoffs between aircraft structure and performance vs AEW antenna type/size/shape/weight must be made and based on a rigorous threat and mission analysis as above
- more development at S-Band frequencies is needed since S-Band is preferable to UHF and L-Band from an ECCM and clutter rejection standpoint
- it is necessary to use maximum practical "core avionics" common to ASW, AEW, navigation, communication, guidance, flight and engine controls, pilot displays and countermeasure equipments
- it is desirable to have maximum possible combined AEW/ASW radar compared to separate installation

V.5.1-FLIGHT CONTROL

Background

- flight related avionics is grouped as follows

    -processor
    -control
    -display
    -guidance
    -navigation
    -communication/identification

- the driving factors which require that the avionic systems for VSTOL be significantly advanced over current systems are:

    -need for excellent flying qualities
    -realistic single pilot workload
    -small ship operations under extreme environmental conditions
    -reliability

- features required by flight control system

    -all digital
    -triply redundant fly-by-wire control
    -integrated electronic cockpit with suitable HUD
    -sophisticated stabilization
    -modular hardware and software
    -fiber-optics multiplexed busses
    -full authority digital engine control

- display systems

    -flight and piloted simulator experience with advanced avionics displays under IFR approach, hover and landing indicate that relatively sophisticated display systems are needed to reduce pilot workload

- further studies are required to optimize the display presentation format for both day and night operations
- advanced flight related avionics can be developed given sufficient engineering support, simulation and flight dynamics testing
- human factors will be very important as pilots and engineers learn together what the most desirable features are

Issues/Problem Areas

- sufficient data are not available to provide a data base for establishing tradeoffs among automated controls, manual controls and display complexity
- needed avionics is many years away from becoming operational
- current diverse programs do not tie together to form one complete system
- a complete set of requirements is not available

V.5.2-AEW RADAR

Background

- AEW mission avionics are dominated by the radar system because of
    - weight
    - size
    - configuration
    - operational flexibility
    - cost
- radar antenna affects entire aircraft design and is the prime driver in creating dissimilarities between the AEW and ASW avionics. . . and to a most significant degree, influences the potential for a common airframe
- close integration between the radar system and aircraft is required

Issues/Problem Areas

- what kind of AEW antenna should be used?
    - rotodome
    - radome with planar phased arrays
    - distributed phased arrays within the aircraft, either planar or conformed to aircraft shape
    - combinations of external "fin" arrays and internal arrays
- unknowns
    - the lightest configuration
    - which one gives the best overall mission effectiveness
    - performance capability
    - technical feasibility
- operational mode must also be determined
    - SAR (Synthetic Aperature Radar)
    - OTH/DCT (Over-The-Horizon Detection, Classification and Targeting)
    - directional agility

- radar weight and size

    -sizing of aircraft for AEW mission depends primarily on radar weight, e.g., the antenna and transmitter make up 80% of the E2-C radar weight
    -studies indicate that 50% weight reduction is needed for the AEW mission
    -reducing weight can be achieved only by revolutionary changes in the antenna and transmitter, e.g., solid state transmitters, imbedded or fin antennas, lightweight power supplies

- commonality

    -as prime (and heaviest) sensor, radar is the key piece of avionics to be optimized
    -tradeoffs must be conducted between AEW and ASW avionics suites
    -can the same radar be used for AEW and ASW?
    -can the ASW aircraft take on some of the AEW missions?
    -what operational changes should be considered?
    -what about cost savings and maintainability?

- critical technology questions

    -type of radar antennae, sensors and processors; internal or external to aircraft
    -independent or common radar for AEW and ASW missions
    -frequency bands: UHF or L- or S-Band electronically agile systems
    -operational modes required: SAR, OTH/DTC, mapping, anti-jam, etc.
    -definitions of threat weapons and ECM and ECCM environments
    -power, weight and cooling limitations for distributed solid state transmitters
    -capability of graphite composite structures for radar electronic transmitters and receptions, and electro-magnetic shielding quality
    -fiber optic coupling devices need development
    -packaging and cooling, in-flight access and malfunction identification for fault correction during maintenance of all electronic modules need development
    -advanced information and operational displays need development

VI. REVIEW OF NAVY VSTOL AIRCRAFT PLANS

The weapon system contribution to surface combatants by helicopters, although providing an improved capability, has been constrained by speed and range considerations. In recognition of this constraint, there have been efforts over the years to develop fixed wing VSTOL aircraft which could be employed from surface combatants. The so-called POGO aircraft, the tail-sitter fighter which was tested during the early 50's, was a first effort at such VSTOL development. The first U.S. military utilization of fixed wing VSTOL aircraft was the Marine Corps AV-8A Harrier aircraft, which has the mission of close-air-support under visual conditions. While representing a significant capability in the Marine Corps inventory, the Harrier's weapon delivery, navigation system, and ordnance carrying capability restrict its use in the performance of most sea-based naval warfare tasks. An improved Harrier, the AV-8B, which will improve the Marine Corps capability to support troops ashore, is in the development stage.

In an effort to pursue the design of high performance VSTOL aircraft, the Navy requirement (in the early 70's) for a combination fighter/attack aircraft of VFAX initially contained a provision that the candidate proposals from industry include a VSTOL version of the basic conventional take-off and landing design. When Congress directed that the VFAX would be a derivative of existing lightweight fighter prototypes, this VSTOL option was no longer possible. The F-18/A-18 was the outgrowth of this competition.

The fact that these early VSTOL aircraft developments did not result in wide-spread full scale development and production stems primarily from their limited performance in comparison with their conventional takeoff and landing counterparts. It is not enough for the military aircraft to be able simply to takeoff and land vertically, the VSTOL military aircraft must be able to perform the required missions, be it air superiority, long range attack or anti-submarine warfare.

In the spring of 1975, the Chief of Naval Operations reviewed the status of VSTOL aircraft and VSTOL ship programs. These programs had as a common feature the fact that they were additive to the programs of land-based patrol aircraft and the carrier-based systems. Moreover, the type of VSTOL aircraft which technology could produce in this acquisition competitive environment were substantially less capable in performance to their conventional aircraft counterpart. It became clear, then, if VSTOL aircraft were to be introduced in any significant role, a new approach had to be taken.

In search of this new approach, the CNO, in April 1975, tasked the Deputy Chief of Naval Operations for Air Warfare to conduct a study which would answer the question: Can VSTOL aircraft be developed which can carry out the naval mission requirements now being fulfilled by manned, sea-based, tactical aircraft? In the spring of 1976, after an extensive exploration, the CNO was advised that although such a development program must be considered high risk, it was the opinion of the aviation technical community within the Navy that it is possible to develop and produce VSTOL aircraft which could fulfill future naval air mission requirements and a <u>transition plan</u> was proposed. This transition plan outlined two areas of concern: first, the overall air capability of the Navy must be maintained in view of the importance of naval aviation to overall Navy mission accomplishments. Secondly, the transition could not involve the development of two duplicative and parallel programs of CTOL and VSTOL aircraft. It was the expense of these parallel programs which had previously limited VSTOL aircraft development. The transition plan was to replace, with a VSTOL aircraft design, each carrier based aircraft now in inventory at the end of its service life.

In looking at the inventory of carrier-based aircraft, two general families of aircraft are apparent. First, there is a group of subsonic long range, long endurance, heavy payload aircraft such as the:

- S-3A Viking Anti-Submarine Warfare (ASW) aircraft
- E-2 Airborne Early Warning (AEW) aircraft
- C-2 Carrier Onboard Delivery (COD) aircraft
- KA-6 Carrier Based Airborne Tanker aircraft
- Search and Rescue (SAR) aircraft
- H-46 Marine Medium Assault Helicopter

These aircraft all commence retiring around 1990. Secondly, there is a group of high performance aircraft such as the:

- AV-8B
- F/A - 18
- A-6
- F-14

which all commence retiring in the mid-1990's or later. These aircraft will have to be replaced on the schedule shown in Figure VI-1, whether with a VSTOL or with a conventional aircraft.

In striving to achieve one of the major goals of naval aviation, namely the reduction in the number of types of aircraft, the Navy envisions the possibility of two basic groups of families of VSTOL aircraft to satisfy most missions at sea. The first grouping, called Type A, includes the subsonic multi-mission aircraft requiring replacement in about 1990. Versions of this type will perform such tasks as anti-submarine warfare/airborne early warning, carrier onboard delivery search and rescue, and the Marine medium assault missions. It is hoped these aircraft will have at least a common airframe design and the same propulsion system with maximum commonality elsewhere a primary objective. A second grouping, Type B, will be supersonic high performance multi-mission aircraft which will perform the fighter and attack missions and replace those high performance aircraft commencing retirement in the mid 1990's.

In line with naval aviation objectives of reduced aircraft types, multi-mission designs have been, and will continue to be, a part of planning regardless of whether VSTOL or conventional. Because of the many tasks fulfilled by carriers, the Navy operates nine different types of aircraft aboard carriers today. In the mid-1980's the number will be seven. In the 1990's the goal is to have three basic types: Type A, Type B, and LAMPS III helicopters.

Although the Navy has made the qualified decision to transition to VSTOL aircraft, alternative courses of action will be reevaluated at each major VSTOL milestone. Because of the broad scope of the VSTOL program, extensive and progressive analysis will be completed at each major decision point if Navy planners (and OSD/Congressional staffs) are to knowledgeably assess the continued validity of the VSTOL approach. In particular, the VSTOL approach will be implemented only if an objective analysis clearly shows that an all VSTOL capability as a total system (aircraft, ships, weapons, C^3, etc.) is more effective than the CTOL system capability it replaces.

During the transition to VSTOL, the Navy is confident that its present sea-based, tactical air capability will not be degraded. The transition plan provides for the replacement of current CTOL models, at the normal expiration of their service life, with VSTOL follow-on aircraft. These VSTOL aircraft will use the current inventory of carriers, augmented by one additional large deck

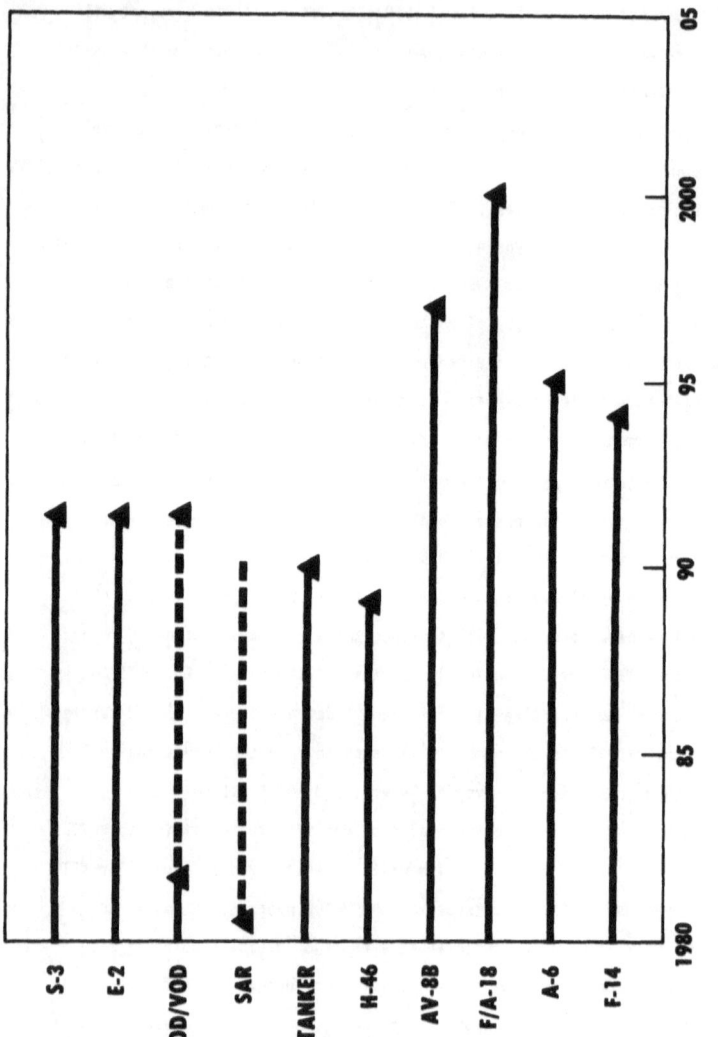

Fig. VI-1. Approximate IOC's for Replacement Aircraft

or equivalent, until an all-VSTOL naval air force makes the pure VSTOL carrier a feasible reality.

The Navy's VSTOL program has the following objectives:

- Transition from CTOL carrier aircraft to VSTOL carrier aircraft for all sea-based, manned, tactical air missions in the U.S. Navy.
- Transition to pure VSTOL carriers when the VSTOL capability has been achieved.
- Phase into an all-VSTOL force by replacing current CTOL aircraft models, at the end of their normal service life, with VSTOL aircraft.
- Maintain the capability of the carrier/naval aviation force at a level not below its current status, while adding fleet capability by providing for VSTOL operations from additional ship types.
- Preclude concurrent CTOL, VSTOL aircraft R&D efforts by phasing the transition as described above.
- Provide alternative courses of action that would allow time to establish plans for introduction of follow-on CTOL aircraft in the early 1990's.

The program is formulated to achieve the advantages of:

- Expanded application of the advantages of manned, tactical aircraft to more platforms in the fleet.
- Improved mission performance of present air-capable surface combatants by providing them with high performance VSTOL aircraft.
- Simplified inventory of naval tactical aircraft with resultant efficiency in training and support.
- Increased flexibility in carrier design.

To develop a VSTOL aircraft that will operate from a broad spectrum of ships requires the application of many new technologies in avionics, airframe, propulsion systems, etc. However, these same technologies will have application to conventional aircraft as well. All of the avionics technologies such as AEW radar are directly applicable. All of the airframe items such as composite materials and lightweight systems have direct application. Advanced propulsion systems have application. The point is, no matter what the final course of future aircraft development (VSTOL or conventional), the VSTOL technology work will have a high rate of applicability for the future Navy.

The Navy plans a phased development program for the Type A VSTOL aircraft weapon system. During the concept formulation phase, now ongoing, industry has been solicited for broad conceptual ideas of operational VSTOL weapons systems. Approximately twenty-three companies have responded to the Navy's request for quotation and information, of which ten were reviewed by NAVAIR. Following an assessment of the information received in response to the unfunded solicitation, a request for proposal will be issued in the summer of 79.

Based upon evaluation of the responses, several contracts for aircraft, engine and avionics conceptual studies will be awarded. These studies will define the operational Type A VSTOL weapon systems, establish performance and cost goals, identify major areas of risk and formulate detailed total development plans.

During the validation phase, design, fabrication, test and evaluation of competitive prototypes, including engines, and hardware demonstration of critical avionics technology will be conducted. Full scale engineering development and follow-on production will lead to the initial operational capability (IOC) in about 1991.

Concept formulation studies for a supersonic high performance VSTOL Type B have just commenced. It is envisioned that a development plan similar to that for Type A will be initiaited in 1980 for an initial operational capability of about 1995.

The order of development, Type A followed by Type B, gives the Navy the replacement aircraft when needed and because of the retirement date spacing of conventional aircraft avoids large concurrent R&D programs.

The Navy will conduct a continuing analysis effort in support of the VSTOL program to delineate the operational utility of the total VSTOL concept. The Navy is analyzing alternate operational concepts, defining utilization of all VSTOL aircraft types and will propose appropriate force mixes. These series of studies will continually address the questions:

- Can VSTOL satisfy the Navy needs?
- Is the concept feasible?
- What is the life cycle cost?

It should be emphasized that the Navy is not committed to an irreversible path. The system of continual program checks is in line with the key decision points of VSTOL and the Navy has structured a program with alternate courses of action as shown in Figure VI-2. There are two key VSTOL Type A decision points. Their relationship to naval aviation planning and alternative courses of action is as follows:

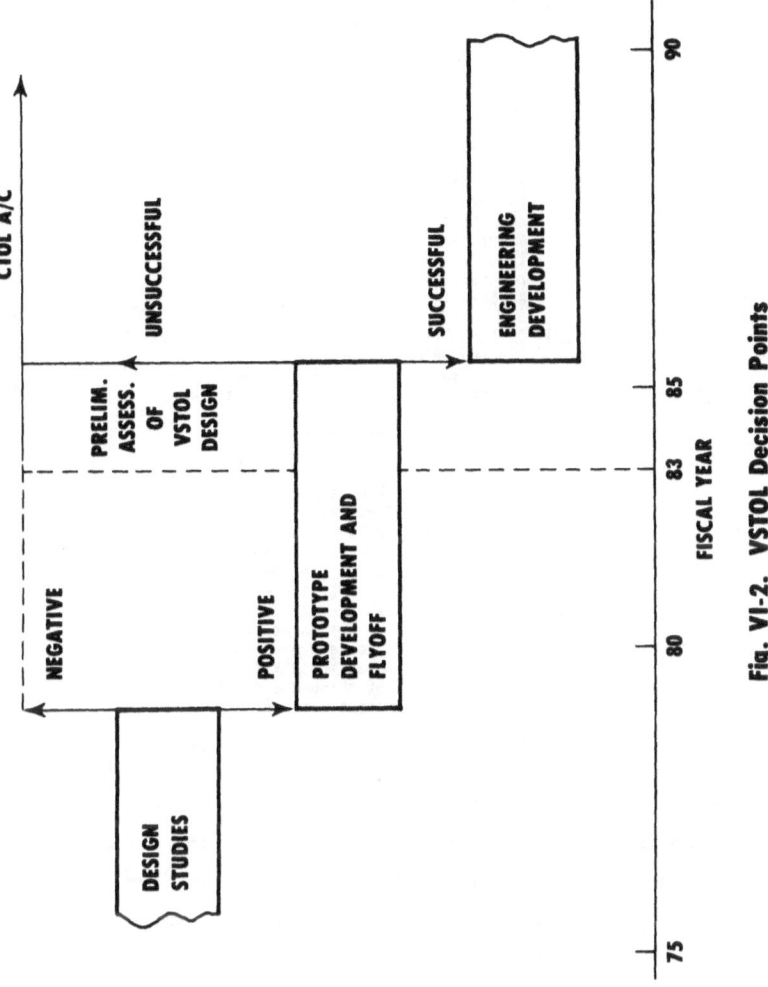

Fig. VI-2. VSTOL Decision Points

(1) If the initial VSTOL conceptual design studies indicate that a VSTOL aircraft cannot satisfy the requirements then no further aircraft design effort will be required until it is time in the 1980's to establish plans for a follow-on conventional aircraft required in the early 1990's. If the design studies are positive, then the Navy will continue with the development program. The competitive prototype fly-off will be completed in 1985 which is another major decision pont. If the prototype fly-off is successful, the program will continue into a normal engineering development.

(2) If the fly-off does not provide convincing evidence to proceed with a VSTOL aircraft development, the program could be redirected toward a conventional multi-mission aircraft to fulfill the same tasks. Since a prototype fly-off would not be warranted due to the relatively low development risk of a subsonic conventional aircraft, it could have an IOC in the early 1990's. Also, as previously mentioned, almost all of the technology and design studies conducted for a VSTOL aircraft could be directly applicable to this conventional aircraft. The point is, the Navy has this period between now and 1985 to thoroughly investigate VSTOL aircraft development through a prototype fly-off without jeopardizing required replacements for carrier-based air in the early 1990's. This has led the Navy to the conclusion that the time is right to pursue the VSTOL concept.

The Navy is proceeding with a detailed appraisal of the VSTOL concept and the preliminary operational and technical studies indicate high potential. As stated earlier, the Navy believes this program offers an effective means for getting the required numbers of manned aircraft at sea to perform the missions of the Navy. Through increased flexibility in staging, VSTOL aircraft may have the potential to expand the combat capabilities of many surface combatants and support ships.

The above approach will ensure that the Navy maintains its ability to support national policy during the transitioning period to VSTOL aircraft. Behind the mainstay of 12 large deployable aircraft carriers, the Navy will maintain the capability to sustain combat operations at sea during this development and transition period. When VSTOL aircraft are introduced into the fleet in the 1990's, they will operate from these large carriers equally as well as conventional aircraft.

VII. VSTOL FUNDING PROFILE

This section contains the best information currently available on VSTOL programs which have been funded by the services since 1950. It should be recognized that much of the VSTOL work has been done as multi-service joint programs. Every effort has been made to identify the individual contribution of each service in these programs. Examples of such programs are the XC-142, X-19A and the X-22.

The Army separated VSTOL data into two categories: supporting research which is directed toward VSTOL aircraft in general and configured research which is directed toward specific VSTOL concepts. Total Army VSTOL funding is $241M.

The Navy did not identify any basic research (6.1) in support of VSTOL. Only exploratory development (6.2) and advanced development (6.3) are provided. The Navy's VSTOL funding profile also excludes the AV-8A Harrier program which was funded from Aircraft Procurement Navy (APN). Total Navy VSTOL funding is $372M.

The Air Force identified $256M in support of VSTOL. This includes basic research, exploratory development, advanced development and a few engineering development (6.4) programs.

In addition to the individual response by each service, the author identified $392M from various and sundry sources dating back to the early 1950s. Therefore, the estimate of total U.S. investment in VSTOL technology and research/flight demonstration vehicles is $1261M.

TABLE VII-1
U.S. ARMY VSTOL PROGRAMS

Supporting Research

FY	RDTE Funds ($K)
50	-
51	-
52	-
53	-
54	-
55	-
56	-
57	1000
58	800
59	2300
60	1300
61	1500
62	1400
63	2300
64	3500
65	5300
66	1600
67	900
68	1000
69	1000
70	5000
71	8000
72	11300
73	11000
74	10300
75	10600
76	10100
77	10900
	$101100 K

U.S. ARMY VSTOL PROGRAMS

Configured Research

Title	PE No	FY Started	FY Terminated	Cost ($K) Thru FY77
Tilt Rotor, Testbed Convertiplane XV-1/XV-3	38-01-017	53	55	15900
Flying Platform, Testbed, VZ-1	38-01-017	55	60	1200
Aerocycle HZ-IDE	-	55	-	175
Wingless Aircraft, Ducted Fan, Testbed	38-01-017	56	59	400
VTOL Rsch. Aircraft Tilt Wing Prop, VZ-2	38-11-003	56	61	3000
Deflected Slipstream Aircraft, VZ-3, Testbed	38-11-004/5	56	59	1700
VZ-4, Rotatable Ducted Fan, Rsch Aircraft	38-11-008	56	60	1900
VZ-5, Vectored Slipstream Aircraft, Testbed	38-11-007	56	56	1700
VZ-6/7/8, Aerial Jeep Rsch Vehicle	38-04-006	58	60	4900
ZV-9, Exp. Rech Vehicle (AVRO)	38-01-016	58	60	4500
S/VTOL Propulsion System	121401A144	62	62	4500
XC-142 VSTOL, Trans Aircraft	63202D153	61	67	29400
X-22A Ducted Prop, Rsch Aircraft	63202D153	62	67	4700
CL-84 Tiltwing Evaluation	63202D155	65	69	300
XV-5A Lift Fan Aircraft	63205D161	61	67	15300
X-19A Tilt Prop, Rsch Aircraft	63202D153	62	66	2400
XV-4A Augmented Jet Aircraft	63205D160	61	65	3600
XV-6A Vectored Thrust VSTOL	63205D162	63	66	20000
High Performance Compound Helicopters	121401A143 & 63211D157	64	66	4200
XV-15 Tilt Rotor Rsch Aircraft	63212DB74 & 63211D157	72	on-going	20000
				$139775 K

TABLE VII-2
U.S. NAVY VSTOL PROGRAMS

Title	PE No	FY Started	FY Terminated	Cost ($K) Thru FY78
Shrouded Propellers	-	50	59	585
XFV-1, Lockheed	-	51	54	9142
XFY-1, Convair	-	51	56	17486
HELO & VTOL A/C Dynamics	-	51	58	444
Low Spd Flt Circul & Flow Viz	-	51	64	736
Int Lift/Prop Veh (Aerodyne)	-	51	59	1088
Single Duct Research Veh	-	54	55	78
Annular Wing Design	-	55	-	15
Propulsion Rotor	-	56	57	799
VTOL Amphib Assault Sys	-	56	58	235
Jet Flap	-	57	-	48
D-188A, Bell	-	57	58	4698
Ducted Prop Tech Anal & Eval	-	57	59	69
Partial Tilted Wing (K-16B)	-	58	61	1865
Lift Thrust Data Analysis	-	58	-	36
VTOL Propeller Prop Sys	-	59	-	50
Ducted Fan Flow Study	-	59	64	311
High Speed Wingless VTOL A/C	-	59	-	30
Kellet Downwash Study	-	60	61	164
Triservice VTOL A/C				
X-22A	62XXX	62	78	30200
X-19	-	61	64	3430
XC-142	-	62	64	28170
Lifting (VTOL) on Ground Effect	-	62	63	42
Flt Path Control Sys	-	63	64	379

U.S. NAVY VSTOL PROGRAMS (CONT'D)

Title	PE No	FY Started	FY Terminated	Cost ($K) Thru FY78
All Mech Stab Aug Sys	-	63	64	237
P.1127 Aircraft	-	63	64	4230
Ducted Fan Improvement	-	63	64	66
VSTOL Naval Mission Anal	-	63	-	69
Adv VTOL Naval A/C Study	-	67	-	95
Adv Carrier Based VSTOL Study	-	68	-	99
Adv VSTOL Fighter Sys Study	-	68	-	98
Light Attack A/C Study	-	69	-	97
Lt Wt VTOL Fighter/Ship Sys Study	-	71	-	98
CL-84 Tests (USN/UK/Can)	62XXX	72	75	1851
VSTOL A/C Development	63257			
TAW Proto (XFV-12A)	W0588	72	on-going	87959
Lift Plus Lift Cruise	W0587	73	76	733
Lift Fan Tech Demo	W0589	74	77	5670
Adv VSTOL (Type A)	W0477	78	on-going	14093
AV-8B Aircraft (Adv)	63211	74	on-going	103653
Adv Prop for VSTOL	63258	73	78	19754
Tilt Wing Design Study	62XXX	73	-	29
VAK-191B Test (USN/FRG)	62XXX	74	76	3474
Jet Induced Effects	62XXX	74	-	109
Ground Footprint	62XXX	74	-	113
VSTOL Tech Assess	62XXX	74	75	75
Selfsurf Sec Power Sys	62XXX	75	77	724
Remote Aug Lift Sys	62XXX	76	77	446

U.S. NAVY VSTOL PROGRAMS (CONT'D)

Title	PE No	FY Started	FY Terminated	Cost ($K) Thru FY78
Var Cycle Eng VSTOL Study	62XXX	76	77	520
X-Wing	62XXX	76	on-going	250
VSTOL HELO Dev (X-Wing)	63203	78	on-going	2100
ABC Flt Test (Tri-Agency)	62XXX	77	on-going	1000
Aerodynamics (VSTOL Support)	62XXX	78	on-going	1742
Propulsion (VSTOL Support)	62XXX	78	on-going	680
Flt Control (VSTOL Support)	62XXX	78	on-going	252
Veh Handling/Serv (VSTOL Support)	62XXX	78	on-going	387
Sensors (Radar, IR)	62XXX	78	on-going	510
Sys Invest	62XXX	78	on-going	1718
Adv Acft Subsys	63217			
Drive Sys Components	W0894	78	on-going	200
Fan Characteristics	W0895	78	on-going	800
Adv AEW Radar	W0446	78	on-going	2229
Avioptics	W0516	78	on-going	300
Digital Fly by Wire	W0687	78	on-going	300
Int Inert Sens Sys	W0886	78	on-going	300
Avionics	63202			
Adv Acft Elect Sys	W0577	75	on-going	5766
Adv Integ Disp Sys	W0579	76	on-going	5780
Aircraft Systems (Adv)	63251			
Lt Wt Hydraulic Sys	W0586	77	on-going	1711
Comp Structure Adv A/C	W0647	77	on-going	1457
				$371874 K

TABLE VII-3
U.S. AIR FORCE VSTOL PROGRAMS

Title	PE/Proj #	FY Started	FY Terminated	Cost ($K) Thru FY78
STOL Tac. Fighter Config. Study	62201/1207	72	-	86
STOL Tac. Fighter Aero. & Control Sys.	"	73	-	38
VSTOL & STOL Tac. Fighter Simulation	"	73	74	44
Ejector-Wing Anal. Pred. Prog. VSTOL Fighter	"	73	-	29
Wind Tun. Test STOL Augmentor Wing Config.	"	73	-	70
Ejector Thrust Aug. Rsch. Veh. for Fighter	"	73	75	106
STOL Tac. Aircraft Invest. Summary Report	"	74	75	27
Ejector Thrust Augmentor Wind Tunnel Test	"	74	-	120
Jt. Navy/AF/MOD VAK-191B Flt. Test Prog.	"	74	-	-
High Lift Multielement Airfoil Anal. & Design	62201/1476	71	72	60
Invest. of Downwash & Trailing Vorticity	"	71	-	15
Jet-Flap Diffuser Invest. Mil. VSTOL Prop.	"	70	72	73
Invest. Ejec. Blown Lift/Cruise Flap Sys.	"	71	74	109
Invest. Double Slotted Rudder	"	72	75	24
Invest. Veh. Generated Turbulence	62201/8219	71	72	96
Estimation of Pilot Rating	"	68	73	18
Anal. Stability & Control Characteristics	"	68	72	33

U.S. AIR FORCE VSTOL PROGRAMS (CONT'D)

Title	PE/Proj #	FY Started	FY Terminated	Cost ($K) Thru FY78
Hand. Qual. Reqts. Piloted VSTOL	62201/8219	69	74	33
Stoppable Rotor Mil. A/C Wing-Rotor Inter.	"	70	72	128
Des. Meth. Tilt Wing Prop. Driven A/C	"	71	72	17
Anal. Rat. Manner Mil VSTOL A/C	"	71	-	56
Appl. Decoupling Meth. in Anal. Design	"	72	73	-
Flt. Path Cont. Resp. in Gusts STOL Mil. Trans.	"	73	76	75
Rev. & Ext. USAF Stability & Control DATCOM	"	73	74	87
USAF Stability & Control DATCOM	"	74	75	73
USAF/NASA Jet Flap Wing WT Test & Anal.	"	74	75	11
Stability & Control DATCOM Revision	"	75	76	68
Flt. Dir. Des. Mil. STOL Aircraft	63205/643A	71	72	76
Pred. Meth. STOL Mil. Trans. Aircraft	"	71	-	100
Des. Criteria Prop/Rotor Mil. Trans. & Combat A/C	"	69	72	1780
Exp. & Anal. Study Mil. VSTOL Flying Qual.	"	71	74	303
Military VSTOL Aircraft	"	71	73	166
STOL Tactical Aircraft Investigation	"	71	73	3955
Control-Display Des. Crit. Future VSTOL A/C	"	71	72	170

U.S. AIR FORCE VSTOL PROGRAMS (CONT'D)

Title	PE/Proj #	FY Started	FY Terminated	Cost ($K) Thru FY78
STOL Transport Thrust Reverser	63205/643A	71	73	198
Mil. Med. STOL Trans. Air Data Meas.	"	72	73	40
Dev. of Reqts. on Direct-Side-Force Control	"	72	74	4
Mil. Med. STOL Trans. Flt. Control Sys.	"	72	73	22
Jet-Flap Wing WT Tests STOL Mil Trans	"	72	74	94
Anal. Mil. & Comm. STOL Design Criteria	"	72	-	99
Direct-Side-Force Ctrl. Mil. STOL Transports	"	72	-	190
Des. Study Common Airframe Demo. Fold & Tilt	"	72	-	300
Dev. Flt. Director for Mil. STOL Aircraft	"	72	73	59
Dev. Pilot Assist Sys. Mil. STOL Aircraft	"	72	73	35
Flow Field Anal. Mil. Turbofan Engine	"	72	-	75
USAF VSTOL Stability & Control DATCOM	"	72	-	50
Propeller Tech. VSTOL Mil. Assault Trans.	64207/69BT	69	73	4907
Stab. & Control Pred. Meth. High Disk Loading	"	69	71	548
Des. Criteria Prop/Rotor Mil. Trans. & Combat A/C	"	69	70	273
Int. Blown Jet-Flap Wing Mod. Mil. STOL Trans.	"	70	-	140

U.S. AIR FORCE VSTOL PROGRAMS (CONT'D)

Title	PE/Proj #	FY Started	FY Terminated	Cost ($K) Thru FY78
Adv. Concepts Turbomachinery	61102/7065	65	74	2039
Math. & Comp. Aspects of Control Sys.	61102/9769	66	72	119
Thrust Augmentation for VSTOL Aircraft	61102/7116	67	70	1119
Fld. Dyn. Energy Trans. Appl. VSTOL Aircraft	"	67	73	1241
Invest. Energy Trans. & Energy Conversion	"	70	73	1003
Design, Fab. & Instal. of Area Ratio 22	"	71	72	395
Exp. Rsch. Fld. Dyn. Energy Conversion	"	73	75	741
Hybrid Prop-Lift Devices for VSTOL Aircraft	"	73	75	534
Factors Affecting Integ. of Augmentors & A/C	"	73	75	263
Anal. Stdy. Aero. Augmentor Wing	"	73	75	318
Mated ARL Augmentor-ATC Diffuser	"	73	-	57
Mass Entrainment by Hypermixing Jets	"	74	75	40
Study of Wall Jets in Diffusers	"	74	-	45
Fluid Dynamic Energy Transfer	"	74	-	588
Research Application Studies	61102/7910	67	71	298
Dev. Simulation Techniques & Programs	62703/6114	67	68	112
VSTOL Aerodynamics	61102/7064	68	73	272

U.S. AIR FORCE VSTOL PROGRAMS (CONT'D)

Title	PE/Proj #	FY Started	FY Terminated	Cost ($K) Thru FY78
VSTOL Oriented Aerodynamics Studies	61102/9560	68	71	820
Lifting Bodies in Low Speed Aerodynamics	61102/9781	68	-	20
Atm. Flt. Aerodynamics & Environment	"	71	75	153
Vibration & Acoustic Envir. for Heli.	62201/1472	68	75	51
Adv. Propulsion Tech. Assess.	-	71	-	50
Mil. Helicopter Flight Director	62201/8226	68	73	46
Helicopter VSTOL Explor. Dev. Vehicle	"	68	76	306
Mil. VTOL Control-Display Sys. Dev.	"	71	75	227
STOL O-2A Flight Test	"	72	73	-
Flt. Test Eval. Low Spd. Sensing Equip.	"	73	-	-
Flt. Test Eval. Doppler Interface Unit	"	75	-	2
Yaw Augmentation Sys. Flt. Test Eval.	"	75	-	1
Takeoff & Landing Crit. Atm. Turbulence	63205/682E	68	71	500
Impact Energy Dissip & Crack Growth in Matls.	62102/7351	69	71	145
Elev. Temp. Titanium Alloys	"	75	-	44

U.S. AIR FORCE VSTOL PROGRAMS (CONT'D)

Title	PE/Proj #	FY Started	FY Terminated	Cost ($K) Thru FY78
Pred. Near-Fld. Noise from Fighter, Trans A/C	62201/1471	69	71	85
Land Gear/Soil Interaction Study	62201/1369	70	72	159
C-130E Main Shock Strut Drop Test Eval.	"	73	-	7
Eval. Ang. Accel. for Adv. Control Sys.	62201/1987	70	74	5
Sub/Hypersonic Air Data Probe Mil. A/C	"	71	76	66
Adv. Eng. Instr. Appl. to Flt/Propulsion	"	73	74	70
Flt/Prop. Ctrl. Coupling & Dyn. Interaction	"	74	76	195
Int. AF & Ind. Flt. Eval. of Military, Etc.	"	74	76	10
Optical Convolution Anemometer	"	75	-	23
Dev. Int. Environ. Control Sys. Designs	62201/6146	70	72	249
Test. Var. Camber Prop. for VSTOL Applic.	62203/3066	70	-	-
Invest. Sand Erosion on Tur. Eng. Compr.	"	70	75	31
Propeller Static Thrust Prediction Methods	"	71	72	82
Wing Pivot Sys. Appl. Analysis	64215/139A	70	75	20
Airload Pred. Avoid. Dyn. & Aeroelastic.	62201/1370	71	72	130
Imprvd. Anal. Soil Model for Predicting, etc.	"	73	74	114

U.S. AIR FORCE VSTOL PROGRAMS (CONT'D)

Title	PE/Proj #	FY Started	FY Terminated	Cost ($K) Thru FY78
Control Tech. Mil. STOL Transports	62201/1986	71	72	9
Landing Approach Turb. Resp. Piloted STOL	"	72	75	4
Study Var. Dia. Rotors	62201/1990	71	-	79
Invest. Methods to Det. Airfoil Pressure	"	71	-	-
Ejector Thrust Aug. for USAF Aircraft	"	72	75	172
Stdy. OVIOA Combat Oper. in SE Asia	62201/4363	71	-	41
Eval. Auto. Ejection Init. Sys. for VTOL	62201/6065	71	74	126
Adv. Fan Stab. & Performance Program	63202/668A	71	73	2894
Unstdy. Motion 2-Diml. Wings IGE	61102/7905	72	76	-
Eff. Rad. Ind. Nausea/Vomiting on Crew	62202/7757	72	77	239
Conceptual Des. Composite VSTOL Tac. Fighter	63211/69CW	72	73	196
Conceptual Design Studies	"	73	74	225
Mod. Biodyn. Effts. Vib. on Pilot Control	61102/2312	73	on-going	241
Rolling Mom. & Side Wash due to Side Slip, etc.	61102/7071	73	-	28

U.S. AIR FORCE VSTOL PROGRAMS (CONT'D)

Title	PE/Proj #	FY Started	FY Terminated	Cost ($K) Thru FY78
Anal. Invest. Med. STOL Trans. Structural Etc.	62201/1368	73	74	175
Dev. Crit. Wide Rge. Transonic Gasdyn. Etc	62201/1426	73	76	524
Concealed Target Detection Sys.	63718/665A	73	-	100
AAA Detection Sys., FAC Aircraft	63743/4316	73	77	400
Four-Diml. Int. Crtl./Disp/Nav. Tech.	62201/2187	74	76	386
Struc. Des. Crit. Spec. & Des. Hdbks.	62201/2401	74	on-going	162
Sonic Fatigue Tests Adv. Mtls. & Structures	"	76	on-going	144
STOL Aircraft Vibration Control	"	77	on-going	150
Acoustics Research	"	77	on-going	75
Mil. Flying Qualities Research	62201/2403	74	on-going	348
Dev. Optical Convolution Velocimeter	"	76	77	50
Multifunction Inertial Ref. Assy. for A/C	"	76	77	298
STOL Assault Trans. Crew Complement Stdy.	"	77	-	42
Dev. of the OCV, Part III	"	77	-	58
Digital Avionics for AMST	62204/2003	74	77	67
Specifications for IDAMST Software	"	76	-	600
DAIS Tech. Anal. & Int. Support	63243/2052	74	76	59

U.S. AIR FORCE VSTOL PROGRAMS (CONT'D)

Title	PE/Proj #	FY Started	FY Terminated	Cost ($K) Thru FY78
Noise & Sonic Fatigue of High Lift Devices	62201/1367	75	76	189
Eval. Adhesive Bonded Structures, etc.	"	75	77	314
YC-15 Interior Noise Meas.	"	75	76	80
Powered Lift Aerodynamics	62201/2404	75	77	162
Jet Flaps for High Lift Airfoils	61102/2307	76	-	40
Resonance Phen. Jet Impng. Sol. Bdry.	"	77	-	35
Hist. Anal. C-130 Life Cycle Costs	63751/1959	76	77	171
Integ. & Appl. Hum. Resp. Tech.	"	77	on-going	493
Crew Workload/Sizing Reqts.	62202/7184	77	-	83
Dev. Noise File 2 & Exp/Update Noise Data Hdbk.	62202/7231	77	-	130
AMST Toxic Gas Studies	62202/7930	77	-	55
Tri-Service VSTOL Dev. (XV-142A)	63206	61	67	90900
VSTOL Engine Dev.	63214	66	71	95100
Consolid. A/C Engr. Lift Engine	63202	62	66	14100
General Electric Lift Engine	"	62	66	15600
VSTOL Propeller Instal. Tech.	"	66	67	1800
Vectored Thrust Engine Study.	"	62	63	700
				$255719 K

VIII. REFERENCES

U.S. Congress, House

Committee on Armed Services, Special Subcommittee on Research and Development, Hearings, Vertical And Short Takeoff And Landing (V/STOL) Aircraft, 88th Congress, 2nd Session, H.A.S.C. No. 68, 1964.

Committee on Armed Services, Report of the Special Subcommittee on Research and Development, Vertical And Short Takeoff And Landing (V/STOL) Aircraft, 88th Congress, 2nd Session, H.A.S.C. No. 69, December 1964.

Committee on Armed Services, Full Committee Consideration of the CVV Program, 95th Congress, 1st Session, H.A.S.C. No. 95-25, May 1977.

Committee on Armed Services, Military Posture and DOD Authorization for Appropriations for FY 78, Hearings, 95th Congress, 1st Session, Washington, U.S. Government Printing Office, 1977. Parts 1, 3 and 4.

U.S. Congress, Senate

Committee on Armed Services, Fiscal Year 1978 Authorization for Military Procurement, Research and Development, . . . , Hearings, 95th Congress, 1st Session, on S. 1210, Washington, U.S. Government Printing Office, 1977. Parts 2, 5, 6, 7, 8 and 9.

Library of Congress, Congressional Research Service

Cooper, B.H., V/STOL Aircraft Development, Issue Brief Number IB78020, March 1978.

Bowen, A., Roles and Missions of Aircraft Carriers in the U.S. Navy: Budgetary and Force Structure Implications, March 17, 1978.

Astronautics and Aeronautics

Navy's Big Order in V/STOL, February 1978, p. 22.

Ross, J.H., Superstrength - Fiber Applications, December 1977, p. 44.

V/STOL Time . . . Maybe, December 1977, p. 8.

Dickert, W.H., Moving V/STOL from Technology to System, December 1977, p. 26.

Few, D.D. and Edenborough, H.K., Tilt-Proprotor, December 1977, p. 28.

Quigley, H.C. and Franklin, J.A., Lift/Cruise Fan VTOL Aircraft, December 1977, p. 32.

Anderson, S.B. and Petersen, R.H., How Good Is Jet Lift VTOL Technology, December 1977, p. 38.

Roberts, L. and Anderson, S.B., Toward a New V/STOL Generation, November 1977, p. 22.

Petersen, F. S., Facing Up to the Military V/STOL Challenge, November 1977, p. 16.

Steiner, J.E., The Timing of Technology for Commercial Transport Aircraft October 1977, p. 42.

Defense Outlook: Smaller, Cheaper, and Slower, October 1977, p. 8.

Eilertson, W.H., A Naval VATOL RPV in Testing, June 1977, p. 30.

Hazen, D.C., V/STOL and the Naval Planner's Dilemma, June 1977, p. 20.

Miller, R. H., A New Era for VTOL? June 1977, p. 18.

Tailsitters Go to Sea, June 1977, p. 12.

Frisch, B., A Civilian Spinoff Shapes Military Aircraft, May 1977, p. 53.

Did the Future Happen in '73?, May 1977, p. 8.

Kurzhals, P.R., New Directions in Civil Avionics, March 1978, p. 38.

Mace, W. D. and Howell, W. E., Integrated Controls for a New Aircraft Generation, March 1978, p. 48.

McIver, D. and Hatfield, J., Coming Cockpit Avionics, March 1978, p. 54.

Aviation Week & Space Technology

Boeing Emphasizes Glass Fiber Rotors, March 20, 1978, p. 57.

Takeoff Procedure Changes Improve Harrier Performance, January 9, 1978, p. 66.

Harrier Ski Jump Takeoff Ramp Tested, December 5, 1977, p. 39.

Navy to Seek Design Proposals for New V/STOLs, November 28, 1977, p. 34.

Advanced Harrier Pushed for Fleet Use, November 28, 1977, p. 55.

X-Wing Aircraft Undergoing Tests, November 28, 1977, p. 63.

Composite Wing Readied for Advanced Harrier, November 14, 1977, p. 21.

AV-8B Pivotal in Naval Aviation Fight, October 31, 1977, p. 18.

V/STOL Design Emphasizes Simplicity, February 7, 1977, p. 17.

Research Effort Seeks Better Reliability, January 31, 1977, p. 118.

Navy Request Stresses Shift to V/STOL, January 24, 1977, p. 18.

Grumman VTOL Aimed at Small-Ship Use, Sentember 20, 1976, p. 15.

Navy Plans Emphasis on V/STOL, March 1, 1976, p. 12.

Additional Sources

Collection of Technical Papers (45), AIAA/NASA Ames V/STOL Conference, Palo Alto, California, June 6-8, 1977.

Aeronautical Vehicle Technology, Technology Coordinating Paper, Office of the Director of Defense Research and Engineering, Washington, D.C., July 1974, (CONFIDENTIAL)

Aircraft Propulsion Technology, Technology Coordinating Paper, Office of the Director of Defense Research and Engineering, Washington, D.C., June 1974, (CONFIDENTIAL)

Technical Area Descriptions, Research, Development, Test and Evaluation, Office of the Director of Defense Research and Engineering, Washington, D.C., April 1977 (SECRET)

Assessment of Sea Based Air Platforms Project Report, Office of the Secretary of the Navy, Department of the Navy, Washington, D.C., February 1978 (CONFIDENTIAL)

O'Rourke, G.G., Our Coming Air-Capable Navy, United States Naval Institute Proceedings, V103, May 1977, p. 90, 92 - 109.

Leopold, J., Designing the Next Aircraft Carriers, United States Naval Institute Proceedings, V103, December 1977, p. 33 - 39.

Taylor, Peter P.W., The Impact of V/STOL on Tactical Air Warfare, Air University Review, p. 65.

Status of the Navy's Vertical Short Takeoff and Landing Aircraft, GAO Report to the Congress, PSAD-78-61, February 23, 1978.

The Technical Basis for a National Civil Aviation Research, Technology and Development (RT&D) Policy, Proceedings of an AIAA Workshop Conference, Crystal City, Virginia, March 1976.

Lind, W.S., A Fleet for the Future, National Defense, September-October 1977, p. 128.

DeLaMater, S.T., Wanted: A Commitment to Make VTOL Work, Vertiflite, September-October 1977.

Slivinsky, C.J., Approaches to Improving Aircraft Efficiency, AIAA 14th Annual Meeting, Washington, D.C., February 1978, No. 78-305.

Petersen, F.S. (Vice Admiral), Keynote Address, AIAA/NASA V/STOL Conference, Palo Alto, California, 6 June 1977.

McMullen, T. H., The Tactical Air Forces - An Outlook, AIAA 14th Annual Meeting, Washington, D.C., February 1978.

VSTOL Technology Assessment, Vol. I - Executive Summary, Vol. II - Final Report, Vol. III - Appendices, Naval Air Systems Command, Washington, D.C., June 1975.

Type A V/STOL, Executive Summary, Lockheed Aircraft Corporation, June 1977 (CONFIDENTIAL)

Technology Assessment of Navy V/STOL Type - A Candidate Weapon Systems Concepts, Riverside Research Institute Contract N00014-77-C-0622, Office of Naval Research, Department of the Navy, Arlington, Virginia, 21 November 1977.

Mow, D.F., V/STOL - Can Navy Needs Be Met?, AIAA 14th Annual Meeting, Washington, D.C., February 1978, No. 78-322.

The Navy's Multimission Carrier Airwing - Can The Mission Be Accomplished With Fewer Resources? GAO Report to the Congress, LCD-77-451, November 16, 1977.

Sea-Based Air Master Study Plan, Chief of Naval Operationa, Washington, D.C., 19 August 1977.

Nutwell, R.M., Gist, D.M., et al, VSTOL Employment in Sea Control, United States Naval War College (SECRET NOFORN)

The Potential Utility of VSTOL Aircraft in Naval Operations, Naval Air Systems Command, Systems Analysis Division, Washington, D.C., Report No. A-503-77-1, March 1977 (SECRET NOFORN)

Type A V/STOL Weapon System, Grumman Aerospace Corp., Grumman Design 698, PDR No. 698-5, January 1978.

V/STOL Fleet Operational Concepts/Requirements Study, Chief of Naval Operations, VFOC/RS Study Doc. 12-1412, 31 August 1977. (SECRET NOFORN)

www.ingramcontent.com/pod-product-compliance
Lightning Source LLC
Chambersburg PA
CBHW030139170426
43199CB00008B/125